# Marswalk One

First Steps on a New Planet

David J. Shayler, Andrew Salmon
and Michael D. Shayler

# Marswalk One

## First Steps on a New Planet

 Springer

Published in association with
**Praxis Publishing**
Chichester, UK

David J. Shayler, FBIS
Astronautical Historian
Astro Info Service
Halesowen
West Midlands
UK

Andrew Salmon, BSc (Hon)
Space Writer
Astronautics News
Smethwick
West Midlands
UK

Michael D. Shayler
Editor and Designer
Astro Info Service
Hall Green
Birmingham
UK

SPRINGER–PRAXIS BOOKS IN SPACE EXPLORATION
SUBJECT *ADVISORY EDITOR*: John Mason B.Sc., M.Sc., Ph.D.

ISBN 1-85233-792-3 Springer Berlin Heidelberg New York

Springer is a part of Springer Science + Business Media (*springeronline.com*)

Library of Congress Control Number: 2005922813

Cover design: Jim Wilkie
Copy editing and graphics processing: R. A. Marriott
Typesetting: BookEns Ltd, Royston, Herts., UK

Printed in Germany on acid-free paper

Printing: Mercedes-Druck, Berlin
Binding: Stein+Lehmann, Berlin

This book is dedicated to the lasting memory of
William H.J. Salmon (1924–1985)
and
Derek J. Shayler (1927–2002)

'Magnificent desolation' – a view of Mars' terrain and horizon, from Spirit, 2004


'Man's first step on Mars will be no less exciting than
Neil Armstrong's first step on the Moon'

Wernher von Braun, from his Manned Mars Landing presentation to the Space
Task Group on 4 August 1969, two weeks after Neil Armstrong took his 'small
step' on the Moon

Other books by David J. Shayler in this series

*Disasters and Accidents in Manned Spaceflight* (2000), ISBN 1-85233-225-5

*Skylab: America's Space Station* (2001), ISBN 1-85233-407-X

*Gemini: Steps to the Moon* (2001), ISBN 1-85233-405-3

*Apollo: The Lost and Forgotten Missions* (2002), ISBN 1-85233-575-0

*Walking in Space* (2004), ISBN 1-85233-710-9

With Rex D. Hall

*The Rocket Men* (2001), ISBN 1-85233-391-X

*Soyuz: A Universal Spacecraft* (2003), ISBN 1-85233-657-9

With Ian A. Moule

*Women in Space: Following Valentina* (2005), ISBN 1-85233-744-3

# Table of contents

# Foreword

On 14 January 2004, President George W. Bush appeared before employees at NASA headquarters and announced 'a new plan to explore space and extend a human presence across our Solar System.' This initiative energises and refocuses NASA on new goals that include manned missions to the Moon by 2015 and a permanent lunar base by 2020, with Mars as an ultimate objective. The President stated: 'We will build new ships to carry men forward into the universe, to gain a new foothold on the Moon, and to prepare for new journeys to worlds beyond our own.'

The schedule presented calls for a new Crew Exploration Vehicle (CEV) to be developed and tested by 2008, for robotic rovers to begin exploring the lunar surface, and for the International Space Station (ISS) to be completed, and the Space Transportation System (Space Shuttle) to be retired by 2010. The CEV is to begin human space flight by 2014, and return to the Moon and establish a lunar base between the years 2015 and 2020. Missions to Mars would commence beyond 2020.

To assist NASA, the President named Edward 'Pete' Aldridge, former Secretary of the Air Force, to be Chairman of the new Space Exploration Commission. The Chairman and his Commission had 120 days to make recommendations to President Bush. Aldridge has stated that priorities will certainly touch on affordability, the talent base at NASA, and contributions by other countries and commercial enterprises. It was no small task for the Commission to tackle the existing bureaucracy and to change the culture of NASA, which had languished in near-Earth orbit for 30 years.

As to affordability, Bruce Murray, former Director of NASA's Jet Propulsion Laboratory, has been quoted as saying: 'It's not a financial issue. NASA's budget doesn't have to increase dramatically for this to take place.' This is probably fortunate, as we are looking at an 'unfunded mandate' with the current state of the federal budget. In addition, until 2010, when the ISS is completed and the Shuttle is retired, 39%, or $6.1 billion, of NASA's current budget is devoted to these activities. In 1961, when President Kennedy mandated that the United States would create a lunar programme called Apollo, funding was not available. However, against much opposition, the President opened the federal coffers and gave NASA carte blanche.

Astronaut Dick Gordon during EVA training for Apollo 15.

Dick Gordon and Jack Schmitt training for Apollo 15. Gordon would have been the thirteenth man to walk on the Moon as Commander of Apollo 18, had it not been cancelled.

Because of this, and the dedication of some 400,000 people, the programme was accomplished.

As with Apollo, the Mars Manned Missions will be preceded by unmanned missions that will provide necessary knowledge. I recall how important it was for Apollo to gain the knowledge provided by the unmanned flights of the Rangers and the Surveyors to the Moon. In fact, Surveyor 3 was a target for my flight on Apollo 12. The equivalent for Mars manned exploration are the flights of Viking, Pathfinder, Beagle 2 (European Space Agency), Mars Global Surveyor, and Russian Mars flights. The surface operations of Sojourner, Spirit and Opportunity have and will provide needed intelligence for future exploration. There will certainly be additional unmanned missions before manned missions to Mars are accomplished. In some regards these relationships somewhat mitigate the heated argument of 'manned vs. unmanned', since we need both to be a vibrant society and nation.

Speaking of debate, I remember one that was crucial to the success of Apollo. In the early 1960s, argument raged about how we would go to the Moon – by direct ascent, *à la Jules Verne*; by Earth orbit rendezvous with spacecraft launched on separate boosters and assembled in Earth orbit before heading out to the lunar surface; or by lunar orbit rendezvous in which the separated vehicles (the Command Service Module and the Lunar Module) would be launched on a single booster and the LM ascent stage would rendezvous in lunar orbit with the CSM after the landing, before the journey home. The direct ascent method was rejected as not being practical. This left Earth orbit rendezvous, which Wernher von Braun favoured, and lunar orbit rendezvous, which the Manned Spacecraft Center, led by Bob Gilruth, favoured. The choice was not made until a high-level management meeting, when someone in attendance reminded those assembled that the mission was to be accomplished 'before the decade of the 1960s was over.' That ended the debate, and lunar orbit rendezvous became the way to accomplish Apollo.

The authors of *Marswalk One* concentrate on what humans will do on the martian surface. Left untreated are the interesting propulsion methods of journeying from the Earth to Mars. Currently available chemical fuels require coasting flight after leaving Earth orbit, and a 7–9-month flight to the vicinity of Mars. This is not very appealing, and other methods and propulsion systems should be available to reduce the transit time to an acceptable level. These propulsion systems include nuclear thermal engines, and ion propulsion that utilises a portable nuclear reactor to heat charged gas. Both of these systems provide acceleration to reduce the transit time and reduce the crew's exposure to zero-gravity flight, which may or may not cause concern. Of great promise is a plasma-propulsion rocket being developed in NASA's laboratories by former astronaut Franklin Chang-Diaz. Although a few years from realisation, by this method the transit time would be reduced to 40 days. This, of course, would be attractive to anyone on a mission to Mars.

*Marswalk One* deals with some of the above, but primarily turns to the science that will be carried out on the martian surface supported by orbital operations – as did the Apollo Command Module (or mother-ship). Hardware to be developed, and the required technology (existing or new), will be explored, as they apply to surface exploration. This book treats subjects such as manned versus unmanned exploration, communications, training, required scientific disciplines, landing sites, martian space suits designed following Apollo and Shuttle EMU experiences, on-site science laboratories, traverses and the duration of such, habitats, lessons learned to be applied to future missions, the return home, the consideration of quarantine after having been gone from Earth for up to two-and-a-half years, and from Mars for the duration of the flight home. Left to the imagination of the reader is the juxtaposition of the human/Mars experience.

I have always said that spaceflight provides one with a different perspective of our planet. While Earth-bound, we all have a certain point of view of our planet either at home, in our cars, on a ship, or during aircraft in flight at 40,000 feet. Flight in Earth orbit provides another perspective, as does flight at lunar distances – which very few humans have experienced. I believe the final perspective will be from Mars, when the Earth, as she is visible, will be that blue dot in the vast void of space.

As in the past, there will be many arguments, discussions and debates regarding President Bush's initiative of a new Crew Exploration Vehicle, the establishment of a lunar base, and eventual missions to Mars after 2020. As an example, it is not intuitively obvious that the CEV can be designed to fly in Earth orbit, travel to the Moon, and handle the task of taking crews to Mars and returning them to Earth. There are those who see 'the Moon as a lunar base' as an objective, and not a stepping stone for martian exploration. Two very prominent individuals share this feeling. One is Louis Friedman, the Executive Director of the Planetary Society, who has been quoted, in referring to the Moon: 'I don't see how it's getting humans ready for another trip to Mars.' Even more outspoken is Robert Zubrin, President of the Mars Society and author of 'The Case for Mars'. Zubrin sees the Moon as a long-term detour, and says that it is 'the same swindle we fell for on the space station ... The schedule is the first red flag. We could go to Mars in six years if we really wanted to.'

To the Moon, or a mission to Mars? There are proponents and detractors for each, or both. NASA's chief long-range planner, Gary Martin, has his work cut out, and he will be a very busy man. I find this very exciting for the future of space exploration. The authors of *Marswalk One* propose much for discussion. Let the debates begin!

Richard F. Gordon
NASA Astronaut, 1963–72
Pilot, Gemini 11
Command Module Pilot, Apollo 12

# Authors' preface

Over the past fifty years, many volumes have been written about the prospects for human exploration of Mars, but there has been very little written about what can be expected from the first crew to achieve the historic landing on the Red Planet. That is the purpose of this volume. The problem with detailing the first landing of humans on Mars is that we have no idea when the landing will take place, how it will be achieved, or how long the crew will remain on the surface – let alone the identity of the crew-members and what they will do when they are there.

Therefore, while other authors have concentrated on how to get to Mars, how to perform extended surface operations over many years, on establishing martian bases, or the theory of terraforming, creating many discussions on the prospect and scope of human exploitation of the fourth planet, we aim to focus on what might take place on the very first landing, and the activities of the first humans to place their boot-prints in the dust of Mars. In many ways the challenge of landing this first crew on Mars will set the stage for the events that follow.

The compilation of this book was all the more difficult because it discusses a programme yet to be defined. Many plans for a Mars mission have been proposed, but no-one has ever put forward a detailed plan of what the first crew will do when they get there, simply because we do not know when or how we will achieve that goal. As there are, at present, no firm plans to mount a human expedition to Mars (although there are plenty of proposals, studies, ideas and dreams), we expect this book to open a discussion focusing on what to include and what to expect on the pioneering landing. Although others may follow after a short interval, what will be the requirements for the initial step onto the planet? Where could they land? What will be the exploration plan? What role will robotic vehicles play in the mission? How far will they travel? How many EVAs could they mount? Will samples be studied *in situ* or on the way home? What types of research will be focused upon by experiments? These are the questions posed for the first landing. Other tasks include deciding upon the type of landing vehicle to support the crew. Will it be self-contained, or will it rely on previously landed robotic vehicles to provide the supplies for the return trip? Will it be rudimentary for short duration, or have facilities for a prolonged surface stay? What will happen in the event injury, death, or a rescue situation? It is clear that, like Apollo aiming for the Moon, until a programme is assigned to reach Mars there will be far more questions than answers. When, at last, we have a firm and sustained commitment to Mars, we can begin to cement the plans for the first landing. Until then, we can only suggest and discuss possibilities.

As this is a book about the future, it is clear that the story will be debated and will evolve until the point where a firm decision is reached, hardware and procedures are defined, and a flight plan is written. This book originated in a discussion between the authors, of the many ideas that have been proposed on how to get to Mars, and when this happens, the sequence of events for setting up a base, and colonising or terraforming the planet. But these are large-scale and long-term goals, and while the very first mission will clearly depend on its timing, the capabilities of the hardware and the final location of the landing site, there still has to be a starting point for surface exploration and a programme for the first crew on Mars. Apollo 11 was the first cautious exploration of the lunar surface, and Apollo 17 ended our first phase of human lunar exploration; but it is expected that the first landing on Mars will be a combination of both caution and exploration, simply because of the distance the crew will have to travel.

The authors consulted many sources to provide a contemporary and informative review of the current (2004) ideas of what the first Mars exploration team will probably be tasked to do. It will inevitability generate further debate and discussion and reveal new information; and this is encouraged, to help generate public awareness of the case for sending humans to Mars, supported by robotic technology and merging the world's space programmes for a common goal.

The authors express a desire to update this work at a suitable point when a programme for surface exploration is more clearly defined. Until then, communication is encouraged though our Marswalk page on the website, http://www.astroinfoservice.co.uk. We hope that a future edition of this book will include a report of the first excursions across the Red Planet, detailing the equipment, procedures, activities and findings of the first humans to explore Mars – when we can clearly identify the date that the first steps on the Red Planet can be identified as Marswalk One.

David J. Shayler

Andrew Salmon
www.midspace.org.uk

Michael D. Shayler
www.astroinfoservice.co.uk                                                   December 2004

# Acknowledgements

Over many years, research into the history, development and operations of various pressure garments has included discussions with, and the assistance of: Joe Kosmo; Jim McBarron; the staff at NASA History Archive, University of Houston at Clear Lake, Rice University, Houston, Texas, and NARA, Fort Worth, Texas, including Janet Kovacevich, Joel Pellerin, Joan Ferry, and Meg Hacker.

For assistance with the scientific aspects of the research, we would like to thank (for robotics) Geoffrey A. Landis, NASA-Glenn RC; Dr Ashley A. Greene, Open University; Dr Andrew Ball, Open University; and Dr Alex Ellery, Surrey Space Centre, University of Surrey; (for geology) Graham Worton, Dudley Museum and Art Gallery; Dr Charles Frankel, Flashline-MARS occupant; Dr Ian A. Crawford, Birkbeck College, University of London; David A. Rothery, Open University; and Jonathan Clatworthy, Lapworth Museum of Geology, University of Birmingham; (for planning, medical and other information) Bo Maxwell, Chairman, UK Mars Society, and MDRS occupant; Dr Michael B. Duke, Colorado School of Mines; Donna Stevens, The Planetary Society; David A. Hardy, space artist.

In addition, the following astronauts and cosmonauts have each offered a personal insight into working in pressure garments, either from the Shuttle on EVA repair, on space stations on long missions, or on the Moon: Joe Allen, Jerry Carr, Gene Cernan, Bill Fisher, Michael Foale, Ed Gibson, Georgi Grechko, Bernard Harris, Jim Irwin, Jack Lousma, Bruce McCandless, Story Musgrave, Bill Pogue, Yuri Romanenko, Jerry Ross, Yuri Usachev, Paul Weitz, and John Young. Additional thanks are also extended to Jerry Carr, for access to his personal library. All illustrations are courtesy NASA, unless otherwise stated.

Thanks also to Dave Portree for his pioneering work on the 'Romance to Reality' Moon and Mars website, and various NASA EVA/Mars exploration documents cited in the bibliography; to Rob Godwin, of Apogee Books; John Charles, of NASA JSC; Lee Saegesser, former NASA historian; the staff of the Zvezda enterprise, Moscow; Andrew Wilson, at ESA; Anders Hanson; and Rex Hall, for allowing unrestricted access to various resources linked to this subject.

Special thanks should be given to US space artist Pat Rawlings, who graciously gave permission to use examples of his excellent artwork and, in particular, for the artwork on the front cover that significantly illustrates the subject of the book.

We must also express sincerest thanks for the continued support of friends and loved ones (especially during ill health): Sue Fairclough, David J. Evetts, Su Fairless, Melanie Cunningham, Ann Parker, MS nurse Amanda Longmore, Iris J. Salmon, Jean Shayler, Ruth Shayler, and Beryl Edge.

The authors also wish to again thank our Project Editor, Bob Marriott, for his continued long hours spent editing and preparing the text, and scanning and processing the illustrations; Jim Wilkie, for his cover design; Arthur and Tina Foulser, at BookEns; the support teams at Praxis and Springer, for the production and promotion of the title; and, of course, Clive Horwood, of Praxis, for his continued support over a long gestation period from the first suggestion of the title to the final publication of the book.

We also look forward to discussing the topics in this book with those who will turn the dream into reality, and hopefully with those who will turn the reality into history – the first Marswalkers. But that is another story, and another book …

# List of illustrations

# Prologue

'By the year 2000 we will undoubtedly have a sizeable operation on the Moon, we will have achieved a manned Mars landing, and it's entirely possible we will have flown with men to the outer planets.' – Wernher von Braun (*c.* 1970).

Following the first Apollo lunar landing in 1969 it was widely expected that authorisation for a programme to explore Mars would soon follow, leading to the first crew landing on Mars by 1980. However, only a year after the Apollo 11 mission, NASA was forced to cut its planned lunar landing programme from ten to just six, having also aborted the Apollo 13 landing due to an in-flight explosion. A variety of domestic and foreign issues had had an impact on the future direction of the US human space programme for the 1970s and beyond. Apollo 17 – the final Apollo mission to the Moon in December 1972 – came home just four years after Apollo 8 had completed the first human voyage to another world. A total of twenty-four American astronauts had ventured to the Moon, and only twelve of these had explored the surface, for three days at most.

Since Apollo 17, all human spaceflight missions from all nations have operated in Earth orbit only. There were many studies and arguments in favour of a return to the Moon and for venturing out to Mars, but nothing ever came of them. In January 2004 – four years after von Braun's latest date for a lunar base and a manned landing on Mars – we finally turned our gaze back to the Moon and out towards Mars.

Therefore, at some yet to be determined day in our near future, we shall once again witness commentary from another world, as we did when we listened to Neil Armstrong on 20 July 1969: 'The hatch is coming open ... Okay, Houston, I'm on the porch.' On the TV screen, a ghostly black-and-white image reveals Neil Armstrong slowly descending the ladder towards the surface. At the bottom of the ladder he drops gently to the landing pad, and then springs back up to the lower rung to test the ascent to the crew compartment. 'OK, I just checked ... Getting back up to that step ... It's not even collapsed too far, but it's adequate to get back up ... It takes a pretty good jump.' Then, as the world holds its breath, history is marked with a paragraph of explanation and a statement of fact: 'I'm at the foot of the ladder ... The LM footpads are only depressed in the surface about one or two inches ... The surface appears to be very, very fine grained, as you get close to it. It is almost like powder ... Now and then very fine. I'm going to step off the lander now'. Armstrong lifts his left foot and, holding on to a lower rung of the ladder, gently but firmly places his boot on the Moon-dust, stating: 'That's one small step for [a] man, one giant leap for mankind.'

Seconds later, Armstrong moved his right boot onto the surface and began to move around in the immediate vicinity of the landing pad, slowly accustomising himself to moving around in the suit in the $\frac{1}{6}$ Earth-gravity. Armstrong had stepped off the LM 109 hrs 24 min 15 sec after leaving the launch pad on Earth. Looking around, he commented on the vista before him, and his first impressions from the surface: 'The surface is fine and powdery ... I can pick it up loosely with my toe [of the boot]. It does adhere in fine layers like powdered charcoal to the sole and sides of my boot. I only go in ... maybe an eighth of an inch, but I can see the footprints of my boot and the tread. There seems to be no difficulty in moving around as we suspected. It's perhaps easier than in the simulations at one-sixth gravity that we performed on the ground. The descent engine did not leave a crater of any size. There is about 1-foot clearance on the ground. We are essentially on a very level place here. It's quite dark here in the shadow, and a little hard for me to see if I have a good footing. I'll work my way over into the sunlight here without looking directly into the Sun. Looking up at the LM, I'm standing directly in the shadow now looking up at the windows, and I can see everything quite clearly.'

One of the first tasks was the collection of a small sample of soil so that, in the event of early termination of the mission, at least a small sample of lunar material would be returned to Earth. Due to the restrictions of the pressure garment, Armstrong was guided by Aldrin from inside the LM and, using a telescopic handle with a scoop attached, gathered the first sample from the Moon and placed it in the leg pocket of his pressure garment: 'This is very interesting. It's a very soft surface, but here and there where I plug with the contingency sample collector, I run into a very hard surface. But it appears to be very cohesive material, of the same sort. I'll get a rock in here ... It has a stark beauty all of its own. It's like much of the high desert of the United States. It's different but it's very pretty out here. Be advised that a lot of rock samples out here, the hard rock samples, have what appear to be vesicles in the surface. I'm sure I could push it [the sample collector] in further, but it's hard for me to bend down further than that.' As Aldrin joins Armstrong on the surface, the two men are struck by the stark contrast and the moonscape before them. Armstrong notes: 'Isn't that something? Magnificent sight down here.' Aldrin sums up a whole panorama of Tranquillity Base in just two words: 'Magnificent desolation.' For a couple of hours on 20/21 July 1969, most of the civilised world was spellbound by the activities of two men a quarter of a million miles away.

It had been eight years and two months since President John F. Kennedy had initiated the quest for the Moon, and less than twelve years since Sputnik was launched to start the space age. But by 20 July 2019, if all goes according to plan, we should see the thirteenth person step on the Moon in time to celebrate the fiftieth anniversary of the famous first Moonwalk by Armstrong and Aldrin. But will we see the first step on Mars by 2029? Will the sixtieth anniversary of the first Moonwalk also coincide with the first EVA on Mars and introduce a new term into spaceflight history: Marswalk One? Whatever the answer to this question, it is clear that the first step on Mars will be nothing like the first step on the Moon.

The Apollo 11 astronauts' descriptions of the lunar terrain were typical of the planned landing area – a relatively flat Mare region chosen in consideration of safety requirements rather than for scientific interest. At Mars, the landing sites of the two Viking landers and the three rovers (Sojourner, Spirit and Opportunity) were quite different, but although each had their specific characteristics, they all had a distinctly 'martian' appearance.

Armstrong's method of exit was to crawl out backwards on his hands and knees to a small porch and down a nine-rung ladder to step off a footpad and onto the lunar surface. The method of stepping onto Mars will very much depend on the final design of the landing vehicle and chosen mode of vehicle exit. The time between leaving Earth and stepping onto the Moon was just over four days, but the difference between leaving Earth and stepping onto Mars will be several months. It is unlikely that the crew will exit only a few hours after landing, and a period of acclimatisation will be required, with surface operations not taking place until several days after landing.

Armstrong noted that the LM descent engine did not produce a crater, because of the low gravity and the absence of an atmosphere, and also because the engine shut off when the contact probe on a landing leg touched the surface. For Viking (1976), the lander engines had a specially designed system of fine multiple-exhaust nozzles, so that there would be very little disturbance of the martian surface as the lander approached. The effect of the first crew-landing will depend on the chosen site and the design of the vehicle braking system. Much of Mars is covered with dust, ranging from large granules of material to particles finer than talcum powder. With the $\frac{1}{3}$ Earth-gravity and the presence of a thin atmosphere, the final approach and landing will be very different from a lunar touch-down. Dust will be disturbed, and will billow around and be blown away from the landing area, leaving a small crater and perhaps a dust cloud that would settle around the landing area. Dust kicked up by the landing engine(s) will cause problems if it settles on any of the solar panels feeding the power systems. Before beginning sampling operations, the crew will have to walk or use a rover to travel to an area of ground undisturbed by their landing and activities.

One of the observations on Apollo was the dramatic contrasts in light and shadow. On Mars, the thin, dusty atmosphere will even out any shadow, so there will not be the crisp shade and sunlit features seen on the Moon. The communications delay between the Apollo crews and Earth was only a couple of seconds, allowing real-time 'live' TV and radio transmissions from the Moon. For a crew on the surfce of Mars, direct communication with the orbital crew would be the most 'local' while in orbit overhead, or via relay satellites deployed in orbit. The orbital craft would not be a 'mission control' as such, but rather an extra set of eyes to oversee the landing area during the risky descent and ascent, and a relay between the surface and Earth (up to twenty minutes away from immediate communications). EVA choreography and mission planning would be another option from orbit, but command and control would be more surface-based, with support from orbit and the final decision resting with mission control on Earth.

The first landing would probably be along the lines of Apollo 12/Surveyor 3, featuring a pin-point landing near an earlier robotic lander. This could be a logistics or habitation module already checked out before the crew departed from Earth, or the return fuel and hardware, depending on the final method of journeying to and from the planet. The crews would also number far more than three (two on the surface and one in orbit) for a week or so in space, and there would more probably be between six and twelve crew-members on a mission of two to three years. Robots would be awaiting them at or very near the landing site, ready for teleoperation by the crew from the lander both before and after the first EVA.

On Apollo 11, the first landing was planned as a single excursion of a couple of hours. Follow-up landings lasted for about 36 hours and included two surface excursions, with the final landing lasting three days with three surface excursions. Although longer surface activities were planned under the Apollo Applications Program, this never materialised. On Mars, the first excursion will probably be quite short, but would be followed by several days and weeks of activity by different crew-members. This will very much depend on the chosen mission profile, hardware design and programme goals. It will certainly be a precursor for even longer and more extensive expeditions in the second and subsequent phases.

Robotic precursor missions paved the way for Apollo, and monitoring from Earth was part of Apollo mission planning. A small programme of crewed missions under the Mercury and Gemini programmes helped refine the techniques and experience to mount Apollo missions of up to two weeks. Long-duration spaceflight operations of more than two weeks essentially came to the forefront after Apollo. For the first landing on Mars, years of research in long-duration spaceflight activities, return missions to the Moon, and intensive robotic activities, will be essential precursors – as will planning for other factors, including minimal solar activity during approach to the planet, avoidance of dust storms at or near to the touch-down areas and back-up sites, and good weather reports from the primary and back-up sites, without hindrance to final approach and landing abort parameters. In short, the landing would have to be chosen at a particular time of year, at the right time of day, in a landing site chosen with minimal dust storms, fog or clouds for a visual or instrument approach. These are just a few of the constraints on a mission that will last between two and three years.

Why? This is the fundamental question behind a human Mars mission! Certainly, science is not the sole rationale for exploring Mars, and other reasons are needed before we send people there. What is the lasting power of this motivation? Even without an answer to this question, humanity is currently searching for the reason to send humans. But, as roboticist Alex Ellery has said: 'Mars is only the first step to opening up the Solar System. It's not the end goal.'

In 1962 – the year after committing America to land on the Moon 'by the end of the decade' – President John F. Kennedy presented another of his famous 'space' speeches to an audience at Rice University, in Houston, Texas: 'We choose to go to the Moon! We choose to go to the Moon in this decade, and do the other things, not because they are easy but because they are hard.' If nothing else, then the same could be said of going to Mars – not because it is easy, but because it is hard ... extremely hard. But once we are there, on the surface, what will we do? How will humans cope with being away from Earth for up to three years, and with living and working in the hazardous environment of a far-off world. That is the subject of this book.

# Destination Mars

As one of the brightest and most recognisable objects in the night sky during its closest approaches to Earth, Mars – the Red Planet – has long attracted human interest. For humans it was at first an object of religious significance, worshipped as a god of war, due to its remarkable rusty colour. Later, we pondered its apparent motion across the sky, puzzling over why it at times appeared to move 'backwards' across the heavens; and then we came to realise that the Earth is not the centre of the Universe and that Mars was simply moving in relation to ourselves around the Sun. After the invention of the telescope, Mars became an obvious target for observation because of its proximity, its colour, and the myth and mystery that surrounded it. Although there was almost nothing to be seen on Venus, we could, at least occasionally, observe Mars in some detail, although it was limited due to the poor definition of early telescopes.

These early observations also gave rise to numerous stories, speculations and theories about the possibility of life on Mars, the presence of seas, rivers and the famous 'canals', and whether or not the dark areas were martian vegetation. American astronomer Percival Lowell certainly fuelled the fire for such speculations in the 1890s when he theorised that the so-called 'canals' he had observed were created by intelligent beings to irrigate the dry equatorial regions of the planet by carrying water from the polar ice-caps. Evidence accumulated later refuted the presence of the 'canals', and proved that the martian environment was not suitable for life as we know it on Earth; but the debate has continued up to the present day, and the most recent missions to Mars have landed with the intention of trying to find evidence that there may have been life at an earlier time in Mars' history, when the environment of the planet was more hospitable than it is now.

It is not the purpose of this book to speculate about the possibility of life on Mars. At the time of writing, nothing definite has been found, but if the current crop of Mars missions finds such evidence, then further scientific investigation will be one of the priorities of any human expedition. Equally, it is not the purpose of this chapter to deal with the myths and misconceptions surrounding Mars. But if we are to send humans to the Red Planet we will need to know as much as possible to ensure that the journey and the stay is as safe as possible for the crew.

The most recent missions to reach Mars may well change some of our current knowledge about the planet, but for the purposes of this chapter we will look briefly at Mars from the point of view of what we know at the end of 2004. One of the assumptions in this book is that at some point humans will set foot on Mars. How this will be achieved, and how we will return, can be left for discussion in other works; but we can make the journey now as we follow the story of the exploratory missions that have already been there. We shall begin with Mars' place in the Solar System.

## TIME AND MOTION STUDY

Mars was instrumental in changing our understanding of how the Solar System works and the realisation that everything within it revolves around the Sun, not the Earth. When religious belief in the heavens was to the fore, it was assumed that the planets moved around the Earth in circular orbits, since the circle was regarded as the perfect form, and nothing less than perfection was appropriate for the heavens. To this end, the astronomer Ptolemy devised his system of planetary motion with the Earth at the centre (geocentric), based on the facts as they were known at the time (around AD 150). At the time there were seven known planets (including Mars), which were said to move around Earth in a series of circles, called deferents. To explain the occasional reverse (or retrograde) motion of the planets, Ptolemy theorised that the seven planets revolved in smaller circles of their own (epicycles) as they travelled around their deferent. Such was the conviction of his theory and the strength of religious belief that Earth was at the centre of the Universe that the Ptolemaic system remained largely unchallenged for about 1,400 years. When the theory was finally overthrown, Mars held the key to the solution.

This monument in Graz, Austria, is dedicated to Johannes Kepler and his discovery of the ellipticity of planetary orbits. (Photograph by Andrew Salmon.)

The breakthrough came with Johannes Kepler, a German astronomer. Continuing the studies of Tycho Brahe – the Danish astronomer to whom he had been an assistant – Kepler came to understand that the orbits of the planets are elliptical, not circular. This realisation helped everything else fit neatly into place, and confirmed the Copernican theory that the Sun, not the Earth, is at the centre of the Solar System (although Copernicus still believed that the orbits of the planets were circular). It was fortunate for Kepler that Brahe's studies had centred on Mars, because the eccentricity of its orbit was the clue he needed to discover the truth. Had he been studying Venus or even Jupiter – the orbits of which are much less eccentric than that of Mars – it would have been more difficult to find the solution.

Mars has an orbital eccentricity of 0.093, compared to 0.017 for Earth and only 0.007 for Venus. There is therefore a substantial difference in its distance from the Sun as it travels around its orbit. The distance of Mars from the Sun ranges from 249,000,000 km at aphelion to 207,000,000 km at perihelion – a difference of some 42,000,000 km, compared with about 5,000,000 km for the Earth. Observations of Mars can therefore be quite variable, as the distance between the two planets at the closest point of their orbits varies each time round. When favourably placed, Mars is one of the brightest and most easily identifiable objects in the sky; but at its faintest it is fainter than the pole star.

**Orbital close-up**

In comparison to the Earth, Mars is about 1 ½ times further away from the Sun. Its orbital period is thus considerably longer and its speed slower, and it takes about 687 Earth days to complete one revolution. This difference also accounts for the apparent 'backwards' motion of Mars across the sky from time to time, which is due to the Earth catching up with and passing Mars as both planets travel around their orbits; and as the two planets reach their closest approach it is always in a different part of their respective orbits. This observation was one proof that Mars could not move around the Earth in a circle at a steady rate.

The point of closest approach between the two planets (opposition, when Mars and the Sun are on opposite sides of the sky) occurs about every 780 Earth days, or 26 months. Using Earth days and weeks for simplicity, Earth will catch up with Mars about every 2 years 7 weeks, at which point Mars will have orbited for 1 year 13 weeks Mars time. Consequently, the distance between the two planets at opposition increases or decreases according to where the opposition occurs around the orbit. For example, the opposition in August 2003 – when Mars approached to a distance of about 56,000,000 km – was one of the closest (and ideal for launching missions to Mars). By the time of the next opposition in late October 2005 the separation distance will be about 69,000,000 km, and at the subsequent opposition Mars will be further away still.

All of this will have an effect on any human mission to Mars, both in terms of when we go and how long we choose to stay there. For a robotic probe on a one-way journey, the calculation is not so critical. It is only necessary to determine the time of a good launch window to make the journey as fast and efficient as possible. Such a fast-track orbit also provides many months at Mars without the Sun interfering with signals.

For a human mission it must also be decided how long to stay on Mars, and it is also necessary to calculate when to re-launch for return to Earth. Three different types of trajectory have been studied for a spacecraft to take humans to Mars. In an opposition type, the spacecraft starts with a transit orbit towards Venus, and the gravity of Venus then slingshots the spacecraft out to Mars. It has a free return time of 14 months, and so any spacecraft can return to Earth 14 months after launch if there is a problem during the early stages of the mission. The other two trajectories take the spacecraft to Mars in a shorter time, but result in a longer return to Earth if there is a problem. Return to Earth from Mars uses one of these three types of trajectory, but not necessarily the same type as used on the way out. The opposition trajectory sends the spacecraft from Mars to Venus to Earth, the fast conjunction trajectory sends the craft back to Earth rapidly, and the Hohmann trajectory sends the craft back to Earth at a leisurely pace.

| Outbound trajectory | Inbound trajectory | Mission length (days) | Time at Mars (days) |
|---|---|---|---|
| Hohmann | Opposition | 700 | 30 |
| Fast conjunction | Opposition | 640 | 30 |
| Hohmann | Hohmann | 980 | 460 |
| Fast conjunction | Hohmann | 920 | 500 |
| Fast conjunction | Fast conjunction | 860 | 540 |

The trajectory determines the amount of time spent in interplanetary space and on Mars. An opposition trajectory reduces the total mission length, but limits the time spent on Mars (1.5–2 years in space, but only a month on the planet). A fast conjunction trajectory, however, minimises the amount of time in space and maximises the amount spent on Mars, and this is the method specified in the NASA Design Reference Mission. Whatever the method, this one physical constraint – the relative positions of the planets – will have a wide-ranging effect on several aspects of any human mission, and on any rescue.

**Home from home**
There are some similarities between the Earth and Mars, many of which make Mars such an attractive target for our first human mission outside the Earth–Moon system. Firstly, both are terrestrial planets (along with Venus and Mercury) consisting of rock and a metallic core, as compared to the gas planets of the outer Solar System. Secondly, despite the length of its year, a single day on Mars (a 'sol') lasts 24 hrs 37 min 22 sec – only slightly longer than a day on Earth. Any human mission to Mars could therefore keep to a similar time schedule to that on Earth, although even a 37-minute difference each day would disrupt the circadian rhythm over time.

Mars also has seasons caused, as on Earth, by the tilt of the planet on its axis and influenced by the distance from the Sun. Mars tilts at an angle of about 24° – almost the same as Earth's – but it tilts in the opposite direction. Earth's orbit is more

This beautiful image of Mars clearly shows the long scar that is Valles Marineris, with three of the huge Tharsis volcanoes to the left. (Courtesy NASA/JPL/British Interplanetary Society.)

circular, and this, coupled with the density of the atmosphere, means that the seasons on Earth are relatively even, although we do notice the change in the Sun's position in the sky.

Due to the eccentricity of Mars' orbit and the lack of any oceans, the difference is much more pronounced. Southern summers on Mars are shorter and hotter than in the north, and the southern winters are longer and colder. Part of the reason for this is that, as with all planets, Mars moves at its fastest when closest to the Sun (as the Sun's gravity 'slingshots' the planet away on its next orbit), and this is when the southern hemisphere is tilted towards the Sun. This will not always be the case, however. Like Earth, the axial tilt of Mars is not constant, because the planet is not perfectly spherical and does not have the stabilising influence of a sizeable moon. But this will not present an immediate problem for any explorers or colonists on Mars. The northern hemisphere will not have its short summers at perihelion for another 25,000 years.

**Observing Mars**
As already mentioned, due to the differences in the orbits of Earth and Mars, conditions for observing the Red Planet vary from opposition to opposition. The most favourable period for observation lasts for about a month, as Mars grows perceptibly larger each day until it begins to recede. Dedicated amateur astronomers, however, follow it for many months, throughout most of the apparition. Mars is a small planet, barely half the size of Earth. It is about twice the size of Earth's Moon, but it is more than 150 times further away. The Earth's atmosphere also interferes with ground-based observation, but so, too, can Mars' atmosphere. Observations made during one of the martian dust-storms can reveal the planet to be almost devoid of features, and such storms can last for a considerable part of the available viewing 'window' of the planet.

Any observations of Mars also depend upon the tilt of the planet with respect to Earth. Until the advent of the space age, the southern hemisphere was much the better mapped because it was tilted towards Earth at perihelion. But regardless of tilt, the dark V-shaped feature of Syrtis Major is still quite visible. Syrtis Major is part of a dark area that extends for a considerable distance around the planet. The dark colour is due to basaltic lava, and although particles of it flake off into fine dust as fast as it is weathered, the prevailing winds blow this dust away to reveal unweathered lava below. It was the movement of this dust and the apparent changing shape of these dark areas that led early observers to believe that there was vegetation on Mars. To the south of it is the vast impact basin Hellas – a circular and almost featureless marking that can often appear as bright as the polar cap, caused when the basin is filled with cloud. The polar caps themselves vary in size seasonally, and are also different from each other because of the varied climate of the two hemispheres. The dark bands that surround the caps are dunes and layers deposited by the dust-storms.

## A RUSTY PLANET

One very obvious query that arises from observing Mars is 'why it is red?' The most commonly accepted answer is that the planet's soil is rich in iron oxide; in effect, the entire planet is 'rusty'. But this then leads to the question of why the mineral is so abundant across the whole planet. It is theorised that the process of oxidation began early in Mars' history when, it is believed, the atmosphere was much denser and surface water was present. Iron-bearing rocks would have been eroded by this flowing water, and the oxygen in the water would then have combined with the iron to create iron oxide, which would then have been dispersed across the planet via rainfall. This theory (and the abundance of iron oxide) has also been cited as proof that substantial surface water once existed on Mars, and with it, possibly life.

Recent martian robotic missions have, however, led to another theory, supported by experiments conducted by Albert Yen of NASA's Jet Propulsion Laboratory, and based principally on the discoveries of the Mars Pathfinder mission in 1997. The mission discovered that the soil on Mars is far richer in oxygen than the rocks, suggesting that some of the planet's iron came from meteorites – which is quite probable, given the level of bombardment that Mars has suffered in the past.

Yen's experiment appears to suggest, however, that the meteoric iron need not have been converted into iron oxide by the presence of water on the planet. In 2000 he experimented on a mineral called labradorite, which is commonly found on Mars. He exposed the sample to Mars-like atmospheric conditions and temperatures, and then bathed the sample in ultraviolet light (to simulate the harsher effects of sunlight on Mars because the planet's ozone layer is negligible). The sample was then analysed for superoxide ions – negatively charged oxygen molecules capable of causing iron to oxidise even when no water is present. The superoxides were indeed present, but they are believed to work too slowly to fully explain the red hue of Mars. Perhaps the colouration of the planet is due to a combination of the action of both water and superoxides.[1]

Mars gradually became more accurately mapped as better telescopes and techniques were developed, and with the advent of the space age and probes sent to the planet, its features were soon mapped and photographed in far greater detail. When Mars is at favourable opposition, most telescopes can distinguish the main features and markings, including the polar caps, and dark areas such as Syrtis Major and the Hellas basin. The use of colour filters in the telescope can also improve contrast and highlight different features: orange for markings, red for maximum contrast of surface features, yellow to brighten deserts, green, blue and blue-green to darken deserts and brighten atmospheric features, and violet to reveal haze and clouds. But it takes larger telescopes (and an absence of dust-storms) to distinguish detail in features such as the great scar of Valles Marineris and volcanoes such as Olympus Mons. Most ground-based observation of Mars is now concerned with phenomena and atmospheric features, because the work of mapping the surface of the planet has mostly been carried out. But the best way to observe Mars is from above the Earth's atmosphere and with a powerful telescope (such as the Hubble Space Telescope), or by taking readings, measurements and photographs in orbit or on the surface.

**Something in the air**
To achieve escape velocity from Earth's gravitational pull, an object needs to travel at about 40,000 kph (11 km/s). Our atmosphere consists of particles which cannot attain this speed, so there is no danger of the atmosphere escaping. However, Mars has a much weaker gravitational pull, and its atmosphere is correspondingly thinner. Surface gravity depends upon a combination of a planet's mass and its diameter: the greater the mass and the smaller the diameter, the greater the surface pull. Mars has only one-tenth of Earth's mass and is only about half its diameter, and its surface gravity is only about 38% that of the Earth, while its escape velocity is only about 5 km/s. Unlike on Earth, most of the atmospheric gases (other than carbon dioxide) escaped, no swathes of vegetation developed, and no oxygen was produced. Mars became the dry desert world that it is today.

Mars' atmosphere primarily consists of carbon dioxide (95%), and is so thin that atmospheric pressure at 'sea level' on Mars is only about 7 millibars, compared with almost 1,000 millibars at sea level on Earth. The atmosphere at martian ground level is therefore no denser than it would be on Earth at 24 km above the ground – almost three times the height of Mount Everest. Humans, therefore, would not be able to breathe it, even if it were pure oxygen. This is another major factor when considering the length of time that we would be able to stay on Mars without some sort of atmospheric processing or sufficient supplies.

Apart from carbon dioxide, Mars' atmosphere also contains traces of nitrogen and argon and negligible amounts of oxygen and other gases such as xenon and krypton. The atmosphere is extremely dry, and with its composition and significantly lower pressure, coupled with the planet's distance from the Sun, the atmosphere does not produce a significant greenhouse effect to noticeably warm the surface of the planet. Despite being 95% carbon dioxide (a known greenhouse gas responsible for global warming on Earth), the atmosphere affects the surface temperature by less than 10°. Mars simply does not retain heat in the atmosphere to keep its surface warm, and temperatures can consequently reach well below $-100°$ C. The Viking missions, moreover, conducted infrared thermal mapping of the planet and revealed temperatures as low as $-133°$ C at the south pole during winter. Temperatures also vary greatly on a daily basis, and Viking recorded a temperature variation of up to 120° in one locality.

The temperature and pressure characteristics of the atmosphere are still sufficient to allow clouds to form (as low-lying fog), particularly over the polar regions, and for winds to attain sufficient speed and strength to create the great martian dust-storms. The dust-storms are caused by the large daily temperature variations between the northern and southern hemispheres and the slight rise in atmospheric pressure during the southern summer, and can escalate into global storms that can last weeks or even months – another major hazard to exploration and colonisation of the planet. One of the longest recorded dust-storms, lasting more than 100 martian days, took place in 1977, during the Viking missions.

The dust is also important in martian atmospheric dynamics. When the dust-storms fill the atmosphere, heat from the Sun is absorbed in the atmosphere rather than at the surface, causing the atmosphere to warm up at the expense of ground conditions. It is akin to the effect of large volcanic eruptions that have been seen on Earth and to what has been suggested as the reason for the extinction of the dinosaurs. A sufficiently large meteor impact on the Earth's surface would throw a tremendous volume of dust and debris into the atmosphere, blocking the sunlight from the surface and cooling surface temperatures to below freezing for long enough to affect all life on Earth. Dust-storms on Mars do not last long enough to have such catastrophic consequences, but they could have a serious impact on any human mission, base or colony.

When the dust settles it presents distinctive markings on the planet's surface. Areas where it settles have a higher albedo (they are brighter) than areas where it does not settle, and there are extensive dust fields in the northern hemisphere and great layers of dust deposits and dunes around both poles.

While current climactic conditions cannot support life and are incapable of sustaining free-flowing water on the surface of Mars, it has been speculated that at some point a warmer climate persisted in earlier martian history. The evidence for this lies in some of the surface formations that can be found on the planet, which, based on our knowledge of how such features were formed on Earth, suggest that water may have once flowed on the surface. If such a warmer climate did exist, it cannot have persisted for long, as the amount of apparent fluvial (water) erosion in such features is quite limited. The existence of preserved impact craters in and around such features also suggests that any water flow and erosion that it may have caused would have been relatively short-lived as the planet's atmosphere dried out.

**On the ground**
One of the most unusual aspects of martian topography is the rather uneven distribution of its geological features. It seems as though two halves of different planets have been put together to form one whole. Most of the southern hemisphere and a portion of the northern hemisphere consists of heavily cratered highlands ranging from 1 to 4 km above what would be considered 'sea level' on Mars. In contrast, the rest of the northern hemisphere (except for the two main volcanic regions of Elysium and Tharsis) is covered with rock-strewn but only partially impacted smooth plains. These are surprisingly flat, sloping only gently (about 0°.05) towards the north pole and mostly lie below the mean surface level.

The highlands cover about 60% of the planet's surface, and the sheer number of large craters suggests that there has been relatively little subsequent alteration or

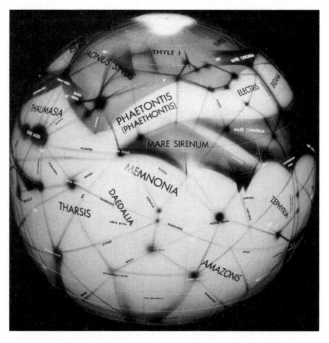

An early three-dimensional reference model of Mars detailing regions and features of the northern plains. (Courtesy British Interplanetary Society.)

disturbance for perhaps millions of years. This area of Mars resembles the surface of the Moon, and it is possible that its history followed a similar course. We know that the Moon suffered sustained heavy bombardment up to about 4 billion years ago, after which impacts occurred far less frequently. Probably the same happened with Mars, although it is not clear why such cratering appears to be concentrated in the highlands. It is possible that Mars suffered a major impact early in its history, and the Hellas basin – 2,300 km in diameter and 8 km deep – is certainly large enough to suggest so.

Although the martian and lunar highlands resemble each other, there are quite significant differences. In particular, extensive smooth areas lie between craters on Mars, probably resulting from volcanic activity at the same time as the planet was being bombarded. There are also branching valleys that look like river valleys on Earth. Both of these features may have had an influence on why the surface of Mars between the craters is smoother than on the Moon. If water, ice or volcanic lava was present in or on the ground at the time of impact, the ejecta from the craters would have been much more liquid and would have flowed across the surface. This is similar to the effect of throwing a pebble into mud, as opposed to the effects of impacts on the lunar surface, which can be likened to throwing a pebble into sand.

The plains of the northern hemisphere are clearly younger than the period of heavy bombardment because the impact craters are few and far between, but there are still significant differences in the apparent age of these plains. The most heavily impacted plains, such as Lunae Planum, were probably formed at a time when the heavy bombardment was coming to an end. The more sparsely cratered plains, on the other hand, are much younger. On the equatorial plains around the volcanic regions of Tharsis and Elysium, the action of the volcanoes has smoothed the terrain and may well have filled in older impact craters under successive lava flows. There are also plains near the north pole which appear to have been caused by a combination of factors, with ice, volcanism, wind and other atmospheric conditions probably all contributing, resulting in a more varied appearance.

**Mountain high**

Among the most impressive features on Mars, particularly in the Tharsis region, are the huge volcanoes. The Tharsis bulge covers about 10% of the planet's surface and is a huge rise in the martian crust about 4,000 km across, with slopes of about 0.2–0°.4 to the north and less to the south where Tharsis merges into the highlands. On the north-west side of Tharsis are the large volcanoes Arsia Mons, Pavonis Mons and Ascraeus Mons, and just beyond its edge is the great Olympus Mons – the tallest volcano on the planet, and, indeed, the tallest known volcano in the Solar System.

Olympus Mons is around 600 km in diameter at its base, and rises more than 24 km above the surrounding terrain – some 2.7 times the height of Mount Everest. Olympus Mons and its three companions are shield volcanoes similar to those of the Hawaiian Islands on Earth. The main difference is their sheer size. The largest of the Hawaiian chain is Mauna Loa – a mere 120 km in diameter at the base and rising to 9 km above the ocean floor. The summit calderas of the martian volcanoes are 10–100 times larger than those on Earth, and the remains of their lava flows are 10–100

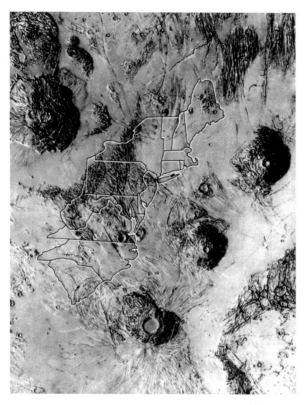

This image gives the scale of the Tharsis volcanoes, with Olympus Mons at top left. Between the volcanoes, the eastern seaboard of the USA is drawn to scale. (Courtesy NASA/JPL/British Interplanetary Society.)

times more extensive. Such figures suggest a large magma chamber and more frequent eruptions than those of the volcanoes on Earth.

The huge size of the martian volcanoes can probably be attributed to the planet's stable and non-moving crust. On Earth, the active lifetime of the Hawaiian volcanoes is relatively short because the plate on which they stand is constantly moving. This eventually cuts the volcano off from its magma source, and it becomes inactive as a new one forms further down the chain. On Mars the volcanoes remained in the same place because there are no moving plates, and they grew bigger from the same source of magma as long as it was available. Given the sheer size and scale of these volcanoes, however, there are relatively few impact craters on their surface, which suggests that the topmost layers of outflow are relatively young. Evidence from the surrounding area indicates that there have been flows for some considerable time on the planet, which suggests that these martian giants are quite old and have had a long active history.

Some volcanoes on Mars are of a type different from these shield giants. Just north of the Hellas basin is a low-lying volcano surrounded by deeply eroded deposits, suggesting that its eruptions were more explosive than those of the Tharsis giants – perhaps of ash rather than lava, similar to Mount St Helens on Earth. Alternatively, the magma could have encountered ice or water as it rose to the surface.

Tharsis has clearly been significant in the evolution of Mars. It is surrounded by a vast system of fractures most probably caused by the sustained pressure of the huge mass of Tharsis on the surrounding crust. On its eastern side there is a huge canyon system and several flood features. Whatever caused the Tharsis bulge to rise, it is clear that it occurred early in the planet's history, because all its visible lava flows, old and young, have moved down its slopes in directions consistent with the current topography of Mars. Despite the size of the volcanoes and the relative youth of their most recent outflows, however, it is improbable that there will be any further eruptions on Mars.

### Grand canyons

To the east of Tharsis is a huge system of canyons called Valles Marineris. It sits like a huge scar across the surface of Mars, and in favourable conditions can be seen from Earth. The feature begins in the Tharsis bulge in an area of fractures called Noctis Labyrinthus, and extends about 5,000 km to the east. Its depth ranges from 2 km at either end to about 7 km in the central section. In comparison, the Grand Canyon, on Earth, stretches a mere 450 km and reaches about 2 km down at its deepest point. If the Valles Marineris were transplanted to the United States it would split the continent in two. It is so long that when one end is well into night, the other is in sunlight, and the subsequent temperature variations send winds along its length.

In the central section, three parallel canyons merge into a chasm more than 600 km wide. The canyon walls have long straight sections, with linear offsets and faceted spurs along the length. These are typical attributes of faulting on Earth, and martian canyons may well have formed in the same way. However, the canyons have also been shaped by other effects. In some places there are deep branching side valleys which appear to have been formed by water erosion – not from rainfall, but by ground-water seepage. In other areas, landslides have widened the canyons. All these different effects have had greater or lesser roles in the formation of the canyons along the length of Valles Marineris. The canyons farther east, for example, are more continuous than those at the western end, and several features, such as tear-shaped islands, exhibit characteristics of erosion by water. At the other end, canyons such as Hebes Chasma are closed off and entirely isolated from the others, suggesting an absence of drainage flow along the length of the feature.

### Channels and valleys

Most of the flood channels on Mars converge in Chryse Planitia and extend for thousands of kilometres until they reach the low-lying plains to the north. There are also channels emerging from the Valles Marineris, which are narrower and deeper as they cross the highlands than they are on the plains.

The channels bear a strong resemblance to terrestrial flood formations, starting and ending suddenly and lacking in tributaries – which suggests that they were formed by tremendous outpourings of water. The martian floods appear to have occurred for two reasons: the sudden release of water pooled in canyons, which may have escaped when one or more of the canyon walls collapsed; or, the sudden release of water under pressure from below the surface, possibly freed as the result of an impact. If sufficient water was trapped under ground, pressures could mount in the same way as in a volcanic eruption. Any impact or other effect that broke through the surface would then release this water – possibly with enough force and speed to collapse the surrounding rock and carry it away. The surface layers would then collapse, creating chaotic terrain. Given the heavy bombardment level and the amount of chaotic terrain on Mars, such eruptions may have happened repeatedly. Another possibility is a volcanic eruption under a glacier. On Earth, this still happens today in Iceland, and causes localised catastrophic flooding.

The martian valley networks are smaller than the flood channels, and have features such as tributaries that appear similar to river valleys on Earth. This suggests that they were formed by the constant erosion of running water rather than by the large flood eruptions. Mars' present conditions and climate would prevent rivers from causing such erosion, because they would freeze too quickly. This has been viewed as good evidence that the climate on Mars was once much warmer. This would not apply to the erosion caused by flooding, as such eruptions would have been so vast that freezing would have a negligible effect even now. However, a warmer climate would not have lasted long, because the valleys did not have time to extend any great distance, nor to cross over and absorb neighbouring smaller valleys or streams. The presence of preserved impact craters across the valleys also suggests that such formations occurred early in Mars' history, and for only a short time.

There are many other contradictory theories about how and when the valleys were formed. Some believe that streams covered by ice and fed by underground springs could have formed them even under present conditions of climate, while others suggest that the valleys were formed by ice rather than flowing water, because at the time that they were formed the Sun's energy would have been insufficient to create the warmer atmosphere required. Determination of the origin of these valleys could yield important clues to the climatic history of Mars and the possibility that life could have evolved.

### Ice caps

As observed from Earth, the polar regions of Mars look very different from the rest of the planet, and appear as bright white areas compared with the rusty regions. There are layered deposits at both poles, extending to about 80° latitude. In the south these deposits lie on the cratered highlands, and in the north they lie on the plains, but both poles have valleys cut into these deposits which show up as distinct swirled patterns.

The polar caps have a definite seasonal cycle, being largest in the winter and shrinking towards the summer, but are not believed to be particularly thick. The south polar cap sometimes extends to 6,000,000 km$^2$, and a thick polar cap of this size would be expected to make the atmosphere wet when melting; but this does not happen, and the polar caps appear to melt quite rapidly with the arrival of warmer weather. It has been found that at each pole there is a quite substantial permanent residual cap which is overlaid by a thinner seasonal cap during the colder months. Due to the low atmospheric pressure there is no flowing water; but there is at least a good reserve of ice, which, for future colonists, makes Mars a better prospect than the sterile Moon.

The polar caps also vary in composition. The northern residual cap consists of water ice, and the southern cap is a mixture of water ice and carbon dioxide ice. The seasonal caps form a covering over the residual caps in the winter, and are composed of carbon dioxide which condenses out of the atmosphere in the autumn and settles over the residual cap.

There are also differences in the composition of the seasonal caps. The northern seasonal cap is much smaller and darker, because it forms at a time when the martian atmosphere contains a considerable amount of dust, while the southern seasonal cap is laid when there is much less dust present. However, the northern residual cap is about 600 km across, while the southern cap is about 400 km across.


## PHOBOS AND DEIMOS

The two moons of Mars were discovered in 1877 by American astronomer Asaph Hall, using the 26-inch Clark refractor at the US Naval Observatory, in the delightfully named area Foggy Bottom, in Washington DC. Hall also chose the names for the two moons from all the suggestions sent to him. In mythology, Phobos was the personification of fear and horror, and Deimos was the personification of dread. Both were sons and attendants of the Greek god of war, Ares (which the Romans called Mars), and so the names were entirely appropriate.

Interestingly, however, the two moons had been 'discovered' well before Hall actually found them. In the seventeenth century it was known that Venus had no moon and that Earth had one, and only four of Jupiter's moons and five of Saturn's moons had been discovered. In a society that placed great significance on numbers it was naturally assumed that Mars must have two moons (and there was, for a time, a search for a 'missing' planet between Mars and Jupiter which was assumed to have three moons). The martian moons were also included in Jonathan Swift's *Gulliver's Travels*. This book is most famed for the little people in the land of Lilliput, but it also included the airborne island of Laputa, whose astronomers told Gulliver about their observations of the two moons of Mars.

Phobos and Deimos are very different from our own familiar Moon. Images returned by Mariner 9 in 1971 revealed them as tiny, dark and very irregular in shape. (Carl Sagan described Phobos as a 'diseased potato'.) The surface of Phobos is irregular in outline and cratered to saturation point, and the formation of a new crater would destroy some of those that already exist. The presence of so many craters implies that the surface is very old – probably dating back to the early days of the Solar System 4.5 billion years ago. The largest crater on Phobos is Stickney (the maiden name of Asaph Hall's wife Angeline), which is so large in comparison to the size of the moon itself that the impact that caused the crater may have been almost severe enough to shatter the moon entirely. Phobos is criss-crossed by 100-metre

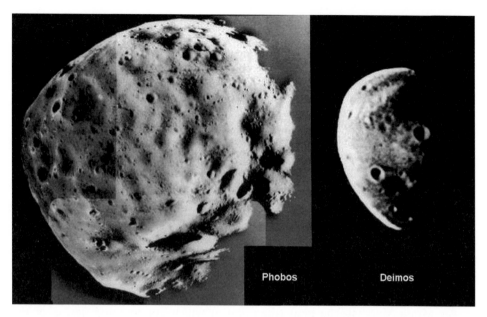

Phobos            Deimos

The two moons of Mars. Phobos has the greater potential to support manned landings on
Mars or as a target in its own right. (Courtesy NASA/JPL/British Interplanetary Society.)

wide, 10-metre deep grooves, and by chains of small craters pock-marking them.
Both of these features may mark fractures induced in the solid body of Phobos by
the impact that created Stickney, but the effects of tidal forces from Mars may also
contribute to these features. Material ejected from Phobos by the impact of
micrometeoroids or larger cratering events does not venture far from the little moon,
and there is probably already a tenuous dust ring of material nearby. The surface
gravity of Phobos (0.001 g) is low enough for material to easily reach the escape
velocity of just 31 km/h, and impact debris probably consists of rock chips and a
flour-like powder. Upon creation of a crater this regolith is ejected straight into
space. It then enters orbit around Mars, but then impacts Phobos and
reaccumulates. Astronauts on Phobos could easily throw objects into space, and
would even fly off into space themselves if they jumped too hard!

Phobos appears to consist of the meteoritic mineral chondrite or even water-rich
carbonaceous chondrite – similar to material in the outer part of the asteroid belt
between Mars and Jupiter (C-type asteroids). The absence of water in the surface
layers of Phobos is a mystery, but it is believed to exist below the surface. The Soviet
spaceprobe Phobos 2 seems to have detected water ions leaking from Phobos, and it
could be a convenient refuelling station for space missions. The water could be
broken down into hydrogen and oxygen for propellants, or used as drinking water
after purification.

Many of the craters on Phobos have raised rims, possibly caused by material that
blew out and piled up around gas vents. This might be caused by the sublimation of
water (changing from solid ice directly to gas), and fits with the theory of a

carbonaceous composition with chemically bound water and black carbon-bearing organic compounds. Some craters have dark patches on their floors and layers of lighter material in the walls.

Phobos is Mars-locked in its rotation (synchronous rotation), and like Earth's Moon as seen from Earth, the same face of Phobos is always seen from a given site on the surface of Mars. On the other hand, the view from Phobos would be spectacular, with Mars encompassing 42° of the sky – about eighty-four times the size of the full Moon as seen from Earth. With the edge of Mars at the horizon, the planet would extend almost half way to the zenith. Phobos is so close to Mars that only about half of a hemisphere can be seen at once (137°), and the surface of Mars would appear to bulge towards the observer. It would be a good site for a remote communications relay station or a weather observatory to provide dust-storm warnings. A temporary base on Phobos could even be used to operate rovers on Mars using telepresence, making use of negligible radio time delays in comparison with remote signals from Earth.

The same tidal forces that have caused Phobos to be Mars-locked are also slowly causing the moon's orbit to decay by about 5 metres per year. In about 40 million years it will approach closer to Mars and be destroyed by gravitational forces as it moves within Mars' Roche limit (the point at which gravitational flexing of the materials inside Phobos becomes too strong and the moon breaks up). It would then form a mini-ring around Mars – a pale rival to the planet Saturn.

Due to the very low gravity of Phobos, a landing would be more like a rendezvous and docking than a touch-down, and anchors might be required. The deep dust pools are a hazard, but astronauts could free-fall from the lander cabin to the surface. Heavy weights would serve to keep the crew on the surface.

Regolith from Deimos can more readily escape into interplanetary space, because

Exploration of the moons of Mars will probably feature manoeuvring units with retention devices rather than surface traverses. (Courtesy Pat Rawlings.)

Phobos lies deeper in the gravity well of Mars. This may explain the difference in appearance of Phobos and Deimos. The latter has more large boulders on its surface. Due to dust infilling, surface features on both moons are subdued. The regolith on the moons is at least a few millimetres deep, and could be as deep as tens of metres. Deimos has no large crater comparable with Stickney, and no grooves. It appears smoother because the regolith is deeper.

The appearance of Mars from Deimos is much less spectacular than the view from Phobos. The planet encompasses 16° of sky – thirty-two times the size of the full Moon. In contrast to Phobos, Deimos is creeping ever further from Mars and will eventually achieve mutual synchronous spin. It will orbit Mars once in the same time that it takes to rotate once and Mars to rotate once. This is also the ultimate fate of Earth's Moon.

Both Phobos and Deimos are possibly asteroids captured by the primordial planet Mars in a form of natural 'aerobraking' by the denser early atmosphere in that location. They bear similarities to objects in the outer asteroid belt. Their colour is under debate (black, white and blue have been mentioned), but they are definitely not the same colour as Mars. Both of them are very irregular in dimensions – Phobos is about 27 × 19 km, and Deimos is about 15 × 11 km – and they each reflect just 6% of incident sunlight (albedo 0.06).

Moonlit nights on Mars bear no comparison to those on Earth. The two moons are too small to make much difference, and their movement across the martian skies is unusual to say the least. Phobos positively scurries across the heavens twice each day, taking just over 4 hours between rising and setting. In contrast, Deimos almost seems to hang at one position, as it takes a leisurely 2.5 days to cross the sky before disappearing for the next three days. Both moons regularly disappear into the shadow of Mars.

Viewed from the surface of Mars, Phobos cannot be seen at all above about 70° latitude, and Deimos cannot be seen above 80° latitude, because both of them orbit so close to the planet around the equator. Phobos is just 2.8 martian radii above the surface of Mars (about 6,000 km) and Deimos lies at seven martian radii (about 17,000 km), while both are only inclined at very small angles to the equator compared with the 7-degree tilt of Earth's Moon. The Roche limit lies at just 2.44 martian radii.

## PROSPECTS FOR LANDING

Mars resembles Earth in many ways. It has an atmosphere (even though it is thin), and has had active volcanoes. Its landscape has been eroded by water and by wind and ice, and it bears rocks and minerals that are familiar to us on Earth. But in many ways the two planets are wholly different. Earth's crust is constantly on the move, forming new volcanoes and mid-ocean ridges, causing earthquakes, and sealing off old volcanoes. But on Mars, areas of eruptions on the surface remained in the same place – which is why the volcanoes grew to such enormous size. The biggest difference, however, is the action of surface water. On Earth, the abundance of water continues to reshape and reform the landscape, while on Mars any erosion caused by free-flowing water was short-lived and has remained intact with the exception of subsequent impacts. This stability has given rise to a spectacularly varied landscape on Mars, with geological features of enormous size. It was into such varied terrain that we had to decide where to send the first Mars probes and landers, without really knowing the true nature of many of these features.

## REFERENCES

1   Albert Yen, 'Ultraviolet Radiation-Induced Alteration of Martian Surface Materials', paper 6076 presented at the Fifth International Conference on Mars, Pasadena, California, July 1999; and Yen, A.S., Kim, S.S., Hecht, M.H., Frant, M.S., Murray, B., 'Evidence that the Reactivity of the Martian Soil is due to Superoxide Ions', *Science*, September 2000.

# History of Mars exploration

The first Solar System explorers were 'robots' such as the Soviet Luna 3, which in 1959 acquired the first images of the Moon's hidden far side, and the American Mariner 2, which flew by Venus in 1962 during the first successful mission to the planets. These craft were designed and built by people, but travelled and operated without a human on board. Solar System exploration has been like this since 1959. Instructions telling a craft what to do are pre-programmed before launch into space, radioed in real time, or uploaded before the destination is reached. (There was also a test of limited artificial intelligence using a computer on the NASA Deep Space 1 comet/asteroid mission, but this was an isolated example.)

## BOLDLY SENDING, NOT BOLDLY GOING

All exploratory missions sent to Mars have, of course, been robotic up to this point, but the history of these missions bears a remarkable similarity to the step-by-step progress of the manned Apollo landings on the Moon. The early Mars probes (at least, the successful ones) were fly-bys; in effect, reconnaissance missions that proved the equipment and the orbital trajectories in much the same way as Gemini and the early Apollo missions (up to Apollo 9) verified the processes and equipment necessary to reach the Moon. These early probes were followed by the first successful Mars orbiter, which returned images of the surface that enabled programme planners to refine the landing sites for the first landers. In effect it did everything except land on the surface (as with Apollo 10), and paved the way for the next stage of exploration. The final stage was the successful landings of the two Viking spacecraft in 1976 (the first of them landing exactly seven years to the day after the Apollo 11 astronauts landed on the Moon). Several of the earlier probes incorporated landers, but these either crashed onto the surface, lost contact soon after landing, or missed the planet altogether. Since the Viking missions, subsequent landers have met with varying fates, but the successful ones have been more sophisticated, carrying more experiments, conducting more science, and even carrying a rover, in much the same way as the later 'super scientific' Apollo missions 15–17.

MAVR (Mars Venera) – a 1960s Soviet proposal for a manned Mars mission, including a fly-by of Venus. (Model at TsNII Mash, Korolev; photograph by Andrew Salmon.)

Over a period of more than forty years, some thirty Mars missions have been attempted, and the relatively small degree of success shows just how difficult it is to reach or land on a target so far away. The much higher degree of success of the Apollo missions to the Moon has been cited as further 'evidence' that the Moon landings were faked, but there are vast differences to be considered. The most obvious one is the difference in the distance between the two targets and the consequences this has for the operation of any mission. Signals to vehicles in the vicinity of the Moon take only seconds to travel to and from Mission Control on Earth. Consequently, any problems can be immediately addressed and corrected. Missions in orbit or landing on Mars are essentially on their own in terms of coping with emergencies or reacting to situations, because signals can take up to 20 minutes to reach Earth. Mission Control can delay or alter the landing, but once the decision is made to land, the robotic vehicle can only follow its programme. No matter how sophisticated the electronics, computers and programming, we have not yet reached the stage at which robotic missions can react to circumstances and 'think through' the options in the same way as a human could react. This combination of human thought and computer programming was another of the factors that made Apollo work, and is likely to affect the degree of success for any human missions to Mars. Controlling robotic missions from a 'mission control' on Phobos would also probably have a greater degree of success because of the short distance that signals would have to travel; but whatever the choice of mission, it is probable that humans will use robotic support on any mission to Mars. They will work as a team. According to roboticist Dr Alex Ellery, of the Surrey Space Centre: 'The robot is going to be an absolutely essential part of any manned mission to Mars.'[1]

**Robotics**

The term 'robotics' can be applied in many different ways:[2]

1   Machines may be entirely under direct human control.
2   Machines may be completely automated, carrying out pre-programmed tasks but with no ability to make decisions themselves.
3   Machines might have a high decision-making ability, operating for long periods without human intervention.
4   Machines could continually interact with humans, and might also have a degree of autonomy (decision-making capability). An autonomous robot can respond to new situations with very little or no guidance from a mission control.

The last category has the most potential for human/robot teams involved in exploration. Another term for this is 'telerobotics', with the robot receiving commands over telecommunications links and with some capability to take action autonomously. For example, a present-day robotic spacecraft such as Mars Odyssey will go to fail-safe mode if it loses navigation lock on guide-stars. Robotic missions are what were also called 'unmanned' or 'planetary' missions, but the ultimate autonomy is artificial intelligence, in which a machine might mimic human intelligence by demonstrating perception, cognisance and reasoning. Robotics also involves mobility and manipulative ability. A combination of robotics (mobility, artificial intelligence and manipulation of robots) and automation provides a machine with greater autonomy (the term sometimes used is automation & robotics – A&R). This would leave humans to carry out the functions for which they are best suited on a mission: reasoning and control for the purpose of discovery.

## EARLY ATTEMPTS

Although the Soviet/Russian space programme has successfully pioneered so many aspects of spaceflight, the Achilles heel of the programme has always been the Red Planet. If the overall percentage of successful Mars missions is low, then individually, the Soviet/Russian success rate is almost nil – which in the main has been due either to long-range communications problems or the failure of launch vehicles. It took four attempts before the first probe successfully left Earth orbit and headed for Mars. It should be mentioned, of course, that this was barely five years after the launch of the very first satellite, Sputnik, in October 1957, and very much still in the infancy of the world's spaceflight programmes. The first two missions – launched on 10 and 14 October 1960 – both ended with an explosion before reaching Earth orbit, while the third, launched on 24 October 1962, at least achieved that goal but exploded before getting any further.

The first successful probe to leave Earth orbit was designated Mars 1, launched on 1 November 1962 (although three days later a fifth Soviet Mars vehicle also exploded in Earth orbit). Mars 1 weighed about 900 kg and carried a variety of instruments and cameras. As with most interplanetary probes it was also equipped with instrumentation to take measurements of the conditions in interplanetary space, recording such phenomena as the solar wind, cosmic radiation, and meteoritic particles. In March 1963, by which time the probe had travelled almost 100 million km from Earth, contact was lost. The vehicle had apparently achieved the correct orbital path to reach Mars, and it is probable that the probe passed the planet at about 200,000 km in June 1963.

### 1965: Mariner 4

By this point, American interplanetary missions had begun, with the first two probes in the Mariner series launched to study Venus. Mariner 1 failed to reach orbit, but Mariner 2 achieved a successful fly-by of our inner neighbour. The Americans then turned their attention to Mars with the next of the Mariner series, launching Mariner 3 on 5 November 1964 and Mariner 4 three weeks later on 28 November. Mariner 3 was lost in orbit around the Sun due to the payload shroud atop the rocket failing to release. This prevented opening of the solar panels and the acquisition of a navigation fix. Mission planners, however, learned from this experience, and modified Mariner 4 to ensure that the problems would not be repeated.

As a result, Mariner 4, weighing in at a little over 250 kg, became the first probe to successfully fly by Mars, to within about 9,900 km, and return transmissions. In July 1965 it sent back twenty-one pictures of Mars, which varied in detail but recorded some seventy martian craters. This prompted comparisons with the surface of the Moon and the thought that Mars was a dead world. Two interesting observations drawn from the Mariner 4 photographs were the first proof of the varied elevation on Mars (one elevation being estimated at about 18,000 metres, although the giant shield volcanoes had not been discovered by this point), and an end to the theory that the dark areas on the planet were due to vegetation. Mariner 4 also provided the first clues about the martian atmosphere, which was achieved by detecting the distortion of the radio signals as the probe passed by and emerged from behind Mars. After the fly-by, Mariner 4 continued to operate for more than two years, and is probably still in orbit around the Sun.

Two days after Mariner 4 was launched, the Soviet Union launched their next attempt: Zond 2. This was another attempted fly-by mission, but its communications failed long before it reached the vicinity of Mars in August 1965. After this there was a lull in Mars missions. The American Gemini programme was coming to an end and the Apollo programme was evolving; but because of the two tragedies of Apollo 204 (Apollo 1) and Soyuz 1 in 1967, both countries put their manned space programmes on hold pending investigations. The next favourable opportunity to reach Mars would not occur until 1969.

### 1969: the Moon and Mars

By the time Neil Armstrong set foot on the Moon in July 1969, four more probes had been launched to Mars. Once again the American missions were successful and the Soviet missions failed. At 3,500 kg each, the two Soviet vehicles were the heaviest so far launched, but both missions suffered booster failures during launch and failed to reach Earth orbit. The American missions continued with Mariner 6 and 7 (Mariner 5 was a successful mission that flew past Venus in October 1967), each of which weighed about 400 kg. Mariner 6 was launched on 24 February 1969, and Mariner 7 on 27 March. The only major concern during these two missions occurred when Mariner 7 suddenly went quiet a few days short of reaching Mars. The silence was due to the explosion of an onboard battery and the resultant tumbling of the spacecraft, but fortunately, signals from Earth were able to reach the probe and it realigned itself on course, although its trajectory had changed slightly. Mariner 7 was also on a faster approach to the planet than was Mariner 6, so although it was launched from Earth some five weeks later, it actually arrived at Mars only a few days behind its sister probe.

Mariner 6 arrived at Mars on 31 July, only a few days after Armstrong and Aldrin walked on the Moon. During its approach it took fifty photographs and a further twenty-five during closest approach at about 3,400 km. Mariner 7, which arrived on 5 August, took ninety-three long-range photographs and thirty-three close-approach photographs at about 3,400 km. The two probes also acquired images of different areas of Mars, with Mariner 6 passing over the equator while Mariner 7 passed further south. They also took atmospheric measurements which confirmed the low atmospheric density, and temperature readings, with Mariner 7 reading the cold polar temperatures very close to the freezing point of carbon dioxide. Of the surface features, Mariner 6 images revealed Olympus Mons as a bright ring, and Mariner 7 scrutinised the Hellas basin. One interesting picture revealed a double crater that was immediately nicknamed the Giant's Footprint. Both probes also searched for the weak magnetic field, but were unable to detect it with the instruments they carried.

After the two probes had flown by, researchers were able to study the returned images and scientific evidence. What they found was not encouraging. The probes had probably covered only about 20% of the planet's surface, and it was unfortunate that they both passed over some of the least interesting terrain. Based on these images, astronomers began to classify Mars as a dull, inert world with no interesting features. Equally, with the determination of the composition of the atmosphere, and the final proof that the dark areas were not vegetation, the possibility of life on Mars receded. The follow-up missions to Mars, however, were too far advanced to be cancelled; but this was fortunate, because perceptions about the planet would be totally changed by the results.

### 1971: Mariner 9, and some success for the USSR

May 1971 was a busy month for launches to Mars. There were five in total: three from the USSR and two American. Of these, two were dismal failures, two were partially successful, and one completely changed our perceptions of the Red Planet. The Americans and the Soviets had different aims for their missions, with the Americans intending to orbit the planet and map as much of the surface as possible, while the Soviets intended to land on the planet and scoop another space 'first' by receiving transmissions direct from the surface.

Each country suffered one of the failures. The first of the two American probes, Mariner 8, launched on 8 May, suffered a failure of its Atlas–Centaur launcher and crashed into the sea. The Soviet failure, Cosmos 419, was launched two days later on 10 May, and at least achieved Earth orbit; but it failed to ignite its upper stage engine

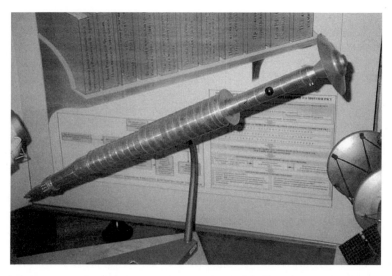

MEK Aelita was a proposal for a Mars mission by the Mishin design bureau in competition with NASA's post-Apollo studies. (Model at TsNII Mash, Korolev; photograph by Andrew Salmon.)

to depart low Earth orbit, and proceeded no further. The designation Cosmos was often given to Soviet spacecraft, in any programme, which failed to achieve the intended goal – part of the Soviet tendency, at that time, to announce each mission only after its was successful.

The other two Soviet missions were Mars 2 and Mars 3. Each weighed about 2,200 kg, and were the heaviest vehicles to reach Mars up to that time. Mars 2 was launched on 19 May, and Mars 3 just over a week later on 28 May. They became the second and third vehicles to successfully orbit Mars (Mariner 9 was actually launched two days later on 30 May, but followed a faster path to Mars and arrived there first). Mars 2 entered orbit on 27 November 1971, shortly after ejecting its landing capsule, which hit Mars to the south-west of the Hellas basin. The lander did not survive the impact, but at least it was the first man-made object to land on Mars. The lander of Mars 3 did not fare much better. The probe arrived in orbit on 2 December 1971, and its capsule came down between Electris and Phaethontis. It began transmitting a picture from the surface, but contact was lost after only 20 seconds and never re-established. The picture received in the USSR showed no detail at all. However, although both landers failed, the two Mars orbiters returned some useful information, mainly concerning atmospheric composition and temperatures.

By the time the first of the two Soviet probes arrived at Mars, Mariner 9 had already been in orbit for about two weeks. It was supposed to operate for about ninety days, but actually sent back more than 7,300 pictures over 350 days in orbit, covering much of the mission of the failed Mariner 8. Mariner 9 carried cameras, an infrared spectrometer, an infrared radiometer and an ultraviolet spectrometer among its experiments, but the pictures it returned initially revealed very little detail. The first of them showed only four spots, which later proved to be the tops of the giant volcanoes of the Tharsis ridge. The reason for this was that Mars was undergoing a planet-wide dust-storm which delayed Mariner's mapping mission by about six weeks. It is probable that this dust-storm had a major effect on the fate of the two Soviet landers. Although Mariner 9 could not obtain detailed photographs at the start of its orbital mission, its other instruments were able to study the dust-storm, which it discovered was lifting dust up to 70 km into the martian atmosphere.

Once the dust had settled, Mariner 9 was able to begin its main objective, and over the course of its transmissions up to October 1972, it took photographs of virtually all of the planet. Its mission also took in the moons Phobos and Deimos, but the most important element was the coverage of the martian surface, which completely changed our view of the planet. Mariner 9 provided the first detailed images of martian topography; and, indeed, the huge canyon system Valles Marineris, which scars the surface, was named after the probe that discovered it. It is fair to say that without the detail uncovered by Mariner 9, the more ambitious Viking missions, launched by the Americans four years later, might not have been undertaken, and would certainly not have been as successful as they proved to be. Mariner 9 eventually ceased transmitting on 27 October 1972; and as far as we know it is probably still orbiting Mars.

### 1974: Red Fleet

By the time of the next launch window in 1973, the Apollo programme had ended and the Americans were left only with Skylab, the joint ASTP flight and the long wait for the Space Shuttle. The Soviets had successfully launched the first space station, Salyut in 1971, having given up on the Moon in favour of long-duration Earth orbital spaceflight. But they had not given up on Mars, despite the failure of their previous landers, and had prepared a fleet of four vehicles, two of which included landers. Riding on their success with probes to Venus, the Soviets were expecting much from their Red Fleet; but once again they had little luck with the Red Planet.

The four probes – designated Mars 4, Mars 5, Mars 6 and Mars 7 – were launched within a three-week window in July and early August 1973. They all launched successfully, and all achieved transit orbit to Mars. It was at the other end of their journey where they failed. Mars 4, launched on 21 July 1973, approached Mars on 10 February 1974, but its braking engine failed to operate and it missed the planet by about 10,000 km. It transmitted a few pictures on the way past, but nothing of any great quality or significance.

The next to arrive was Mars 5, which was launched on 25 July 1973 and arrived on 12 February 1974. This was by far the most successful of the four, and successfully achieved Mars orbit after its braking engine worked. It transmitted some pictures showing craters and river beds, but with nothing like the quality of Mariner 9. Sadly, it ceased transmitting after only a few days.

The remaining pair carried the landers. Mars 6 was launched on 5 August 1973 and arrived at Mars on 12 March 1974. The lander successfully separated and, at least initially, followed a good descent trajectory towards the surface. Rocket braking was used, and the main parachute deployed for about 150 seconds; but then the lander lost contact. Presumably, the lander made a soft landing on the surface, but contact was never regained. The Mars 7 lander was a complete failure. Launched on 9 August 1973, it also followed a faster track to Mars, arriving three days before Mars 6 on 9 March 1974. However, the lander either separated from the orbiter prematurely or there was some failure with its braking or attitude control, because it missed the planet by about 1,100 km and headed off into a solar orbit of its own. After this group of costly failures, it would be some fifteen years before the Soviets ventured back to Mars, and it was left to the Americans to make the first fully successful landings.

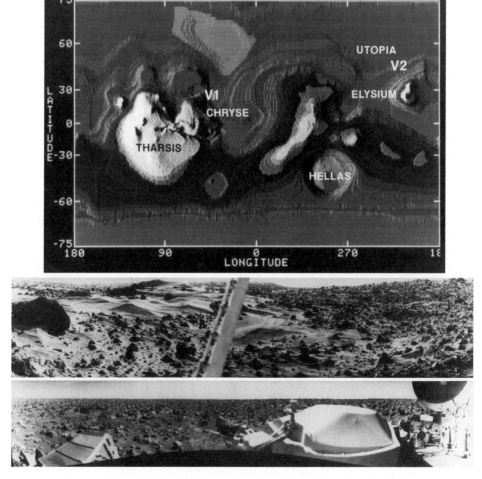

The landing sites for the two Viking missions were chosen using data from Mariner 9. Had NASA realised just how rough they were, neither landing site would have been used. (Courtesy NASA/JPL/British Interplanetary Society.)

**1976: Viking invasion**
The sites for the landings of the two Viking missions had been chosen carefully using the data from Mariner 9. The landings were the next logical stage in exploring Mars after the mapping programme, because information was required about the surface, including soil sample analysis, atmospheric composition data, and some clue as to whether the surface was sterile or capable of supporting life. The two missions were targeted at Chryse and Cydonia – both of which were relatively low-lying and were expected to be smooth. They were also both situated at the end of the drainage system associated with the Tharsis volcanoes, and it was therefore a very favourable location for the detection of any moisture on the planet. As with two of the previous Soviet probes, both Viking missions consisted of a lander and an orbiter, with the orbiter conducting its own investigations and also acting as a communications relay between Earth and the lander. Weighing about 1,500 kg each, these were, at the time, the heaviest probes sent by the Americans.

Viking 1 was launched on 20 August 1975, with Viking 2 following almost three weeks later on 9 September. Both launches and transit journeys were made without any major incident, and Viking 1 arrived at Mars on 19 June 1976, followed by Viking 2 on 7 August. Due to advances in imaging technology, far-encounter pictures taken by both probes surpassed anything received from Mariner 9. The first task for Viking 1 was to study the proposed landing area in Chryse, and the first photographs taken of the area showed it to be much rougher than expected – to such an extent, moreover, that the eventual landing site was changed twice, still in Chryse but further west. One of the main concerns for this first American landing mission was to ensure that the lander did not descend onto a rock that might puncture the spacecraft and render it useless. In the event, it touched down within 30 km of the revised impact point; but it had certainly been lucky, as it was less than 10 metres from a boulder large enough to badly damage the craft.

The very first picture transmitted after touch-down showed a rock-strewn landscape. The lander's legs had not penetrated far into the dusty soil, indicating a solid surface, but the overall impression was of a barren rocky desert. Later, the lander sent a colour picture showing the rusty colour of the soil, but one major surprise was the colour of the sky. Once the transmitted pictures were corrected, the sky was found to be pink – caused, apparently, by fine dust particles suspended in the atmosphere and absorbing the blue light from the Sun. Both Viking landers carried an array of experiments, including temperature, pressure and wind sensors mounted on a telescopic boom, cameras, mass spectrometers to analyse the atmosphere, X-ray fluorescence spectrometers to determine elemental composition, experiments including gas chromatography and pyrolysis for molecular analysis, and a three-axis seismometer. They were also equipped with a biological laboratory containing a gas-exchange experiment, a labelled-release experiment, and a pyrolytic release experiment, all designed to test for signs of life on Mars. The only one of these to fail was the seismometer, which could not be activated because the cage used to protect it during the descent would not detach.

Viking 2 encountered a problem after the lander separated from the orbiter. There was a slight power failure in the orbiter that caused the spacecraft to drift off its main communications link. This needed to be corrected to prevent all the information collected by the lander being lost, because the orbiter would have failed to pick up the lander's signals to relay them back to Earth. Fortunately, low-power communications worked to ensure that no data was permanently lost, but the problem was not completely cured until after the lander touched down.

The other major event was a complete change of landing site. Like Viking 1, the probe was set to study its landing area in Cydonia prior to landing, and, as for its predecessor, this landing site also appeared to be too rough. Eventually, however, it was decided to aim for a different landing site in Utopia Planitia, some 6,500 km north-east of Viking 1. This site was expected to be fairly flat, but instead of being a gently rolling landscape of sand-dunes and dust, Viking 2 pictures showed a landscape similar to Chryse, with rocks strewn everywhere. Experiments and investigations eventually showed that Utopia is a volcanic landscape. The seismometer on Viking 2 worked, but surprisingly detected no marsquakes apart from the results of tremours produced by the mechanisms of the lander's experiments. The first marsquake was recorded in February 1977.

Up to 1982, both landers transmitted more than 4,500 pictures from the surface, and the orbiters transmitted more than 51,500 images and mapped 97% of the surface down to 300-metre resolution. These would be the last Mars missions for more than ten years, and the last American missions for more than fifteen years. In another parallel to the Apollo Moon landings, at the moment of our greatest success on Mars, we stopped going there.

### 1989: a different target

After the disappointment of the Red Fleet, it was a long time before the Soviets sent any new missions towards Mars; and when they did, it was in cooperation with international partners including France, Germany, Great Britain and America. The missions were targeted at one of the martian moons. Phobos 1 and Phobos 2 were launched within a week of each other on 7 and 12 July 1988. Each had an ambitious programme of experiments to study both Phobos and Mars, and each carried a lander to land on the surface of Phobos. Phobos 2 also carried a smaller lander equipped with a spring system that would enable it to 'hop' over the surface. Each probe weighed about 2,600 kg.

Once again, however, the Soviets had little luck. On 2 September 1988, contact with Phobos 1 was lost on the way to Mars, caused by human error in uploading an incorrect command to the vehicle. Effectively, the probe was told to turn itself off, and onboard equipment failed to override the command. Phobos 2, however, obtained some good images on the way to Mars, and reached the planet on 29 January 1989. But although operations initially proceeded according to plan, contact was lost with the probe on 27 March, before the most important phase of the mission had begun. This time the failure was attributed to a fault in the onboard computer. It was not a total failure, however, as it detected the magnetic field of Mars; but once again the Soviets had little to show for their efforts. These would also be the last Mars missions under the banner of the Soviet Union.

**The 1990s: mixed fortunes**

The protracted interval between Soviet missions was exceeded by the interval between the successful Viking missions and the next American attempt: some seventeen years. Mars Observer – launched on 25 September 1992 (after a delay caused by a hurricane) – was crammed with sophisticated experiments and equipment. The experiments included a magnetometer, a laser altimeter, a thermal emission spectrometer and a gamma-ray spectrometer. Its objective was a mapping and information-gathering exercise, particularly of the martian atmosphere, and it was planned to operate for at least a full martian year. Launch and transit to Mars proceeded according to plan, and Mars Observer obtained its first image, showing an almost cloudless planet, on 26 July 1993. Shortly afterwards, however, the mission came to an end. As part of the braking manoeuvre, the radio link was shut down so that the fuel tanks could be pressurised. The link should have been recovered some 9 minutes later, but contact was never regained. Some contingency had been made for this, and Mars Observer should have reorientated itself and switched to a back-up receiver if it received no signals from Earth for five days, but the last signal came back at 0100 UT on 22 August, and by the end of September NASA had given up on it.

The next mission to Mars was undertaken by Russia, and once again it was an ambitious programme. Mars-96 carried some twenty experiments covering atmospheric, surface and core research, and the package included two small landers targeted at Arcadia and Amazonis. Two further landers were targeted at Acidalia and Utopia, but these were penetrators that would plummet into the martian surface. The ground stations were expected to operate for at least a martian year, and the orbiter would continue mapping and studying the planet as well as acting as a relay for the landers. However, none of the experiments worked because Mars-96 never even left Earth orbit. The fourth stage of the Proton launcher began a long engine firing, but shut down far too early. The probe then separated from the booster and fired its own engine, but it was not sufficiently powerful to achieve escape velocity. Launched on 16 November 1996, the probe re-entered Earth's atmosphere and burned up just two days later on 18 November (preceded by parts of the Proton booster on 17 November). To date (2004) this is the last spacecraft that the Soviets/Russians have attempted to send to Mars.

The previous missions had been failures, but the two American missions launched at the end of 1996 were among the most successful interplanetary missions of all. Mars Global Surveyor, launched on 7 November, was already on the way to Mars at the time of the Mars-96 disaster. On 4 December it was followed by Mars Pathfinder, which travelled in a more economical path and arrived at Mars first, two months earlier than Mars Surveyor in July 1997.

## MARS PATHFINDER AND MARS GLOBAL SURVEYOR

Mars Pathfinder, weighing only 460 kg, was relatively small. Its lander was shaped like a pyramid, with three triangular sides designed to open out after landing to expose the package of experiments. This package included atmospheric entry science, and long-range and close-up imaging. Uniquely up to this point, the lander also carried a small roving vehicle called Sojourner. This rover carried its own camera and solar panels, but its main objective was analysis of the rocks at the landing site, and for this it carried an Alpha-Proton X-Ray Spectrometer (APXS). Sojourner was able to approach and physically contact a target rock and bombard it with alpha particles from the APXS. The behaviour of these particles could then be analysed, counted and measured to determine the abundance of different elements in the rocks. Sojourner was expected to have an operational lifetime of about a week.

One other unique aspect of the Mars Pathfinder mission was its method of landing. Apart from the heat shield and the parachute, Mars Pathfinder was protected by airbags which would allow it to bounce to a halt on the surface. It was an unusual method, but it worked. The airbags were inflated immediately prior to jettison of the main parachute, and the lander dropped the last hundred feet, hitting the surface at speed and coming to a halt after some seventeen bounces, the first of which carried it back up to a height of more than 500 feet. After its final touch-down it rolled along the surface for about a minute, but it eventually came to a rest intact and upright. The only problem was that one of the deflated airbags came to rest on top of the lander, but this was soon resolved. Almost immediately, Mars Pathfinder began transmitting images of the landing site in Ares Vallis – a reasonably flat area, but with plenty of rocks for Sojourner to investigate.

Both Mars Pathfinder and Sojourner operated successfully for much longer than anticipated, with the last clear transmission being received on 27 September 1997. By

An artist's impression of the Mars Pathfinder landing, showing the air bags deflated and the Sojourner rover already starting its chemical studies of the rocks. (Courtesy British Interplanetary Society.)

then, however, Mars Pathfinder had sent back some 16,000 images, and Sojourner the composition of the rocks and the 'fog' clouds. The first rock examined (given the name Barnacle Bill) was found to be similar to terrestrial andesites, which are volcanic and can be found on Earth in places like Iceland. Another rock (Yogi) was found to be different – less rich in silicon, and basaltic. The rocks studied by Sojourner were found to be similar to those found at the Viking sites, and rounded pebbles were seen both on the ground and in the hollows of some of the rocks, indicating the presence, at some stage, of running water. For a mission put together quickly and on a small budget, Mars Pathfinder was a great success.

Mars Global Surveyor was equally successful. It arrived at Mars in September 1997, at about the time that Mars Pathfinder was coming to the end of its life, but its mission was completely different. Mars Global Surveyor was an orbiter-only probe, with instrumentation designed to map the planet. Its operational life was expected to be about two years, during which time it would explore all aspects of Mars, including the atmosphere, surface topography, mineral distribution and the planet's magnetic properties. It carried several experiments, each designed for specific areas of its programme of investigation. The onboard Mars Orbital Camera (MOC) was designed to produce a daily 'weather satellite'-like image of Mars, and the Mars Orbital Laser Altimeter (MOLA) measured surface relief to produce elevation maps of greater accuracy than had previously been possible. The Thermal Emission Spectrometer (TES) was designed to analyse infrared surface radiation and measure temperatures and the composition of rocks and soil. Finally, the Magnetometer Electron Reflectometer (MER) was included to search for any evidence of a magnetic field. It was this instrument that produced the first major discovery on 15 September, barely four days after orbital insertion.

Primary mapping was scheduled to begin in March 1998, but a malfunction in one of the solar panels delayed it for about a year. However, the MOC provided high quality images from the start, and has produced a great deal of information about the geology of, for example, the canyons in the Noctis Labyrinthus area. The MOLA has produced a detailed topographical map of Mars, and has been used in conjunction with gravitational data to construct models of the thickness of the martian crust. At the time of writing, Mars Global Surveyor has had its mission extended, and is expected to be active until at least December 2006.

**A new player**
With the launch of Japan's Planet-B on 4 July 1998, a third country entered the arena. Planet-B was scheduled to enter orbit around Mars in October 1999 on a mission to investigate the planet's atmosphere. It was soon renamed Nozomi (Hope), but this proved to be a misnomer, as the mission was plagued with problems. Its trajectory to Mars was originally planned to include fly-bys of both Earth and the Moon, but during the Earth fly-by on 20 December 1998 a thruster malfunctioned and left the probe with insufficient velocity to reach Mars. A couple of correction manoeuvres also burned more propellant than intended, and the trajectory had to be recalculated. The revised trajectory involved two more Earth fly-bys, but this rescheduled the mission for arrival at Mars in early 2004 – some five years behind

schedule. Furthermore, in May 2002 a solar flare damaged the probe's electronics, and although the subsequent Earth fly-bys enabled the probe to eventually head for Mars, the operators were unable to correct a short circuit (caused by the solar flare) to allow the probe to enter Mars orbit. Early in 2004 it flew past the planet on its orbit around the Sun.

**American setbacks**
By the time the Japanese had calculated the revised trajectory for Nozomi, two more American probes were already on their way to Mars. Mars Global Surveyor had been the first of NASA's Mars Surveyor series of small, low-cost spacecraft for sustained exploration of Mars, and these two new probes were the next in the series. Unfortunately, they were nothing like as successful.

On 11 December 1998, Mars Climate Orbiter (MCO) was launched on a mission to investigate the martian atmosphere, climate and surface conditions, and to serve as a communications relay for the second probe, Mars Polar Lander (MPL), launched on 3 January 1999. Once Mars Polar Lander had completed its mission, Mars Climate Orbiter would have continued on its own mission to monitor the atmosphere and surface to form a picture of climate changes. Scientists hoped to learn about the planet's climatic history, and possibly about buried water reserves. At the end of its mapping mission, Mars Climate Orbiter was also expected to act as a communications relay for subsequent missions in the series. It arrived in the vicinity of Mars and began its orbital manoeuvres on 23 September 1999, but it failed to re-establish contact after passing behind the planet. An investigation into the problem revealed a fatal error in the commands sent to the probe from Earth. This was caused by confusion between NASA and the manufacturer Lockheed-Martin. NASA's specification for thruster firings was metric, but Lockheed was working in imperial, and the mistake was not detected until it was too late. Discrepancies in small trajectory manoeuvres built up during the flight to Mars, and this caused the orbiter to miss its projected orbit at 140 km altitude, fall into the martian atmosphere at about 57 km, and burn up.

Mars Polar Lander was equally unsuccessful. Its mission was to deploy a lander and two penetrators – called Amundsen and Scott – on the surface of Mars to gather data about the planet's past and present water resources. The mission was targeted at the southern polar carbon dioxide ice-cap (which had never been studied), recording meteorological conditions, analysing samples, and taking images of the terrain. The two penetrators would have impacted the surface about 100 km from the main lander to obtain small samples of subsurface soil. However, none of this actually occurred. Although contact with the probe was lost (as planned) on 3 December 1999, about 6 minutes prior to atmospheric entry, it was never regained.

NASA attempted to find the probe using Mars Global Surveyor, but without success. The investigation into this failure suggested that it was probably caused by erroneous signals during the deployment of the lander's legs. The signals indicated that the lander had already reached the surface when it was still descending. Consequently, the main engine shut down prematurely, and the lander plummeted to the surface and crashed.

**Mars Odyssey**

Mars Odyssey – another mission in the Mars Surveyor series – was an attempt to refly the experiments from the Mars Observer orbiter failure. It was launched on a Delta 2 rocket, from Cape Canaveral, on 7 April 2001, and was a great success. The original plan was to use aerocapture to enter orbit around Mars, where, upon arrival, a dive into the atmosphere would reduce most of the velocity before the spacecraft entered orbit around the planet. This was later changed to an upper stage for a solid motor firing, followed by aerobraking using a solar panel, with several months spent dipping into the upper atmosphere – the technique used and proven by Mars Global Surveyor. Aerobraking followed arrival in October 2001, and science began in earnest in February 2002. Surface maps of mineral composition were produced using infrared and gamma-ray spectrometers. Some areas of Mars had never seen water (they were rich in olivine), while others had minerals (such as haematite) indicating the presence of water. An infrared imaging spectrometer and a visible-light camera made more maps showing surface composition and topography, even at night, and gamma-ray and neutron spectrometers were a key component of NASA's 'follow the water' mantra. They revealed vast amounts of subsurface hydrogen (probably in the form of water – $H_2O$). A radiation monitor (MARIE) from the Johnson Space Center collected information during the cruise to Mars and upon arrival, and found that radiation levels at Mars were 2–3 times higher than in low Earth orbit. MARIE proved troublesome at first (as had a similar instrument on the International Space Station), and it was silenced for good by a large eruption of solar material that reached Mars in October 2003. Mars Odyssey acted as a data relay for craft on the surface of Mars (the two Mars Exploration Rovers), and during attempts to contact the UK/ESA Beagle 2 lander. The relay task will continue into at least 2005.

**Another new kid on the block**

The Mars Express mission is the first European Space Agency 'flexible mission', developed at relatively short notice and at low cost through maximum reuse of components from, for example, the Rosetta cometary probe. The spark was the loss of many valuable European scientific instruments on the Russian-led Mars-96 mission. Mars Express was launched on a Russian Soyuz rocket, with a Fregat upper stage, from the Baikonur Cosmodrome on 2 June 2003. Apart from the reflight of the four Mars-96 instruments, the payload included the 40-metre long Marsis subsurface radar mapper, a monitor of the solar wind at Mars (ASPERA), and the Beagle 2 lander. Beagle 2 was the brainchild of Professor Colin Pillinger of the Open University, and included instruments from Hong Kong (a rock drill), Germany and Russia (the PLUTO 'Mole'). Mars Express suffered a 30% power loss after launch, which affected full scientific operations, but it safely entered orbit around Mars on Christmas Day 2003. The first European planetary mission was the Giotto probe to Halley's comet in 1986, but Mars Express marked Europe's entry into both Mars exploration and planetary exploration. Beagle 2 was successfully released from its carrier spacecraft on 19 December 2003, but nothing more was heard from this ambitious and innovative craft, despite numerous searches and attempts to contact it.

**Rovers hit the ground 'on the bounce'**

The rovers Spirit and Opportunity (their names were chosen by American school-children following a NASA/LEGO® toy company competition) were launched by Delta 2 rockets from Cape Canaveral on 10 June and 7 July 2003 and, attached to their cruise stages, reached Mars on 4 and 25 January 2004 respectively. They proceeded to land on the surface immediately upon arrival, without entering Mars orbit. The entry, descent and landing system for the Mars Exploration Rovers (MER) used a heat shield, followed by a parachute, braking retro-rockets, and finally inflatable airbags to cushion the landing. Unlike Mars Pathfinder, all of the Mars Exploration Rovers' scientific instruments were carried on the rovers themselves. Spirit landed at Gusev Crater – a possible ancient lake-bed – and Opportunity landed in Meridiani Planum, which had large deposits of grey haematite mineral, possibly formed under wet conditions.

The rovers acted as 'field geologists' on Mars, using a suite of scientific instruments deployed at each location that was visited. Cameras took visible-light and infrared images, scientific instruments studied rock samples after the Rock Abrasion Tool removed their weathered surface layers, and a thermal emission spectrometer, similar to that on Mars Global Surveyor in orbit, provided infrared 'ground truth'. They also carried a Mossbaueur spectrometer to measure iron-bearing materials in the rock, an Alpha Particle X-ray Spectrometer to investigate the composition of the rock, and a Microscopic Imager to closely examine the rock grains and 'soil'.

The rovers had a planned life of three months, and were set to traverse about 1–10 km around the landing sites during that time. The factors which limited their life were the accumulation of dust on the solar panels, and day/night extremes of temperature on the rover components. Small pellets of non-weapons-grade plutonium provided heating by radioactive decay (Radioisotope Heater Units), and instructions on where to go each day were prepared at the Jet Propulsion Laboratory. Results from the scientific instruments were sent to the Mars Odyssey orbiter and then on to the Deep Space Network on Earth.

Both rovers have overcome several problems and computer reboots to explore a range of terrain, and remained operative throughout 2004. In September 2004, NASA granted a funding extension of six months, with the provision that the rovers continue to operate. Both rovers have investigated very different terrain in the course of their missions. Opportunity ventured deep into the 130-metre wide, 20-metre deep crater Endurance, while Spirit struggled up the slopes of Husband Hill (named after the Commander of the fatal 2003 *Columbia* Shuttle flight, STS-107). Its struggles proved worthwhile, as it encountered bed-rock containing high levels of bromine, sulphur and chlorine, which seemed to have been highly altered by water. Shortly afterwards, both rovers were effectively put into hibernation as the martian winter solstice approached (September in the Earth calendar), and the amount of energy the solar panels could draw from the Sun was reduced. At the same time, radio communications were interrupted as Mars reappeared from behind the Sun and moved away from conjunction. With the Sun and Mars in the same part of the sky (as seen from Earth), radio communications were badly disrupted for about twelve days. The two rovers have, however, survived the harsh conditions, and are set to continue their missions well into 2005.

## FUTURE PLANS

Several more missions are planned for the next few Mars launch windows, and with the January 2004 announcement of plans to send humans to Mars, these missions may well change over the years as we learn more about Mars and develop any programme to go there ourselves. The results from the current and future missions will have an influence over what comes next, and in turn, over where we land and what we do when humans finally arrive. The data gathered from these robotic missions will need to be thoroughly analysed to choose the best landing site for the historic first human mission.

### Safe landing sites

The return of samples to Earth by robots will be essential in advance of human missions. Whenever people might go, robots will go beforehand. The Apollo missions to the Moon were preceded by the Ranger hard landers, the Surveyor soft landers and the Lunar Orbiters; but there were time constraints. Adhering to John F. Kennedy's original announcement in 1961, NASA had to land men on the Moon by the end of the decade, so it needed to be achieved as fast as possible. With Mars, however, time will not be a factor, although maintaining political and public support could be a problem if it takes too long or costs too much. In the case of Mars, samples will be returned before humans arrive to verify that the materials of the martian surface will not cause problems if inhaled inside the habitat. There is no point in conducting an EVA, doffing the suit in the airlock and handling martian surface samples, only to find that breathing in regolithic material is hazardous.[3] Another reason for sending robotic precursor missions would be to test the bearing strength of the surface to ensure that a given area is safe for a landing. There are many different types of terrain on the surface of Mars, including the so-called 'stealth' terrain, which does not easily reflect radar signals from the surface. Here the material is a very fine, light dust aggregate – facetiously called 'foo-foo dust' – that may have zero or almost zero bearing strength. It is the martian equivalent of quicksand.

### Where do we go first?

The surface of Mars is as diverse as the land surface of Earth, and covers about the same area. A landing site for the first human mission should have a wide variety of terrain within 1,000 km – a site where science could be carried out for fifty years or so. One option is to then send later missions to this same location to build up surface infrastructure, and the whole surface of Mars could be explored from one landing site.[4] Such exploration would use telepresence operator astronauts located at the landing site, linked by communications satellites in martian orbit to robotic rovers exploring the whole of the martian surface. Samples could then be retrieved from each exploration site and returned to the initial landing site for study and later return to Earth if sufficiently interesting. Long-range pressurised rovers could also be sent out so that the human crew could explore the surrounding regions. A possible site for such exploration is the Coprates Quadrangle, which incorporates the Tharsis volcanoes, Valles Marineris, the Margaritifer Sinus region of ancient cratered terrain, including several paleolakes, Kasei Vallis (the site of run-off channels), the outflow channels region north-west of Valles Marineris, the northern plains near Chryse, and the Viking 1 landing site.[5]

Before the first human landing, however, a considerable amount of data would need to be gathered:[2]

- Three-dimensional mapping of the surface to 10-cm resolution for the landing area and the zones to be covered by crewed rovers and robots.
- The mechanical properties of the regolith.
- Dust adhesion properties of airborne dust and regolith.
- The total absorbed radiation dose in tissue or an equivalent material on the surface of Mars (charged particles and neutrons).
- The presence and concentration of highly toxic hexavalent chromium ions in the regolith and airborne dust.
- The potential corrosive effects of the regolith and airborne dust in a humidified environment.
- Determination of whether organic carbon is present at or above a life detection threshold value in the zone of the landing site.

This information would be acquired by orbital mapping, robotic sample return missions and/or *in situ* measurements specific to the landing site. Some martian weather systems – including dust-storms and dust-devils – would also pose a possible risk to landings, surface modules, rovers, and astronauts on EVA. Environmental factors (such as weather) have been blamed for the Soviet Mars and the Beagle 2 lander failures, although this is not certain, and could be attributed to something as simple as parachute system failures.

**Landing sites**

A tremendous amount of effort goes into choosing a landing site for robotic missions to Mars, just as the most important contribution of the US Geological Survey to the Apollo programme was its choice of lunar landing sites.[6] This would be even more critical for human missions to Mars. The members of a landing site working group have to balance the two conflicting demands of safety and potential science results, and a staggering number of factors must be considered.[7] A safe place to land might be quite boring scientifically, but an interesting site might prove too hazardous for a landing. In the case of the first human mission, the safety and return considerations may well override the science, as it did with Apollo 11.

In the days of the Viking landings there were two steps in the process. First the landing site was selected, and then it was certified. The selection began years before the landing, with a large number of candidate sites reduced during reviews at working group meetings, which were often contentious. Today the process revolves

This meteorite called EETA 79001 was found in Antarctica. It is believed to have come from Mars because it contains a small amount of gas that is similar to the martian atmosphere. (Courtesy NASA/British Interplanetary Society.)

around the initial site selection without 'certification',[8] because there is sufficient visual material available from orbital mapping. The approach used for Mars Pathfinder and later robotic landers was to aim to understand the surface properties at the landing site, using remote sensing, and models based on the results.

First and foremost in any site selection are the engineering constraints arising from the design of the spacecraft landing systems. Scientific objectives are factored in once safety is assured. Scientist Matt Golombek best summarised the situation when helping to choose a landing site for the 1997 Mars Pathfinder mission: 'You're flinging this expensive piece of hardware at Mars, and if you do not land safely you do not get any science.'[9] The engineering constraints evolve with the design of the lander itself, but site selection is a convergence between engineers who know the capabilities and limitations of the machines that they are building, and scientists who determine the scientific worth of the areas accessible to the lander. If lander systems or the scientific outpost to be established rely on solar power, then the landing site must be at a latitude exposed to direct sunlight for most of the day; that is, it must be near the equator. Too high a latitude will be bad for two reasons: the Sun will be too low on the horizon, and temperatures will be lower. This will place more demand on power systems to keep other systems warm.

The terrain must also be taken into consideration. The landing site must be below a certain altitude to provide the lander with sufficient time for ballistic slowing and for the parachutes to open properly and slow the lander before the retro-rockets fire. For Mars Pathfinder, this took 55 seconds, so the maximum altitude had to be 1.3 km below the reference altitude used for the MERs.[10]

Wind speeds experienced during descent also have to be within certain limits to ensure proper functioning of the parachutes. For the Viking landers, any site that might experience wind speeds of more than 70 m/s was ruled out, even though the craft could, in theory, land in winds of up to 240 kph. Analysis of wind-induced streak-marks recorded on photographs can be used to avoid high wind areas, and the MERs examined both atmospheric boundary layer winds (up to 5–10 km) and near-surface wind effects (the lowest 5 km) such as wind shear. This brought into play local topography (canyons, for example, funnel the winds), the season (surface pressure may be slightly higher in the spring), and the time of day or night (the atmosphere is less dense during the day) for the landing.

The angle of the terrain at the landing site also needs to be suitable for a safe landing. This was set at a 10-degree limit for the (cancelled) Mars Surveyor 2001, and was 19° for the Viking landers in 1976. Under these circumstances the gradients on sand-dunes could pose a hazard, as well as masking the effects of rocks. Equally, landing on the edge of a mesa poses a threat, as the retro-rockets might be triggered to fire by a radar altimeter, only to suffer from a last-minute drift over the edge of the mesa, where they run out of braking propellant during the unexpectedly long descent to solid ground. On the other hand, the retro-rockets might fire too late if the lander drifts towards a steep rise of land. The 2003 MERs used atmospheric models produced using Mars Global Surveyor TES infrared data and topographical data from the MOLA laser altimeter.[10]

At the landing site there must be a minimal number of rocks tall enough to cause a crash of the lander. The rules for the Viking landers dictated 'free of rocks and other hazards greater than 22 cm in height,' while the cancelled Mars Surveyor 2001 mission allowed for 'a less than 1% chance of landing on a rock greater than 33 cm high.' In 2003 the MERs classed any boulders of 0.5-metre height or 1-metre diameter as 'hazardous'.[11] For engineers, the ideal planet to land on would be as smooth as a billiard ball,[12] and the eventual Viking landing sites would not have been chosen at all if there had been images available showing the surface rocks.

The thick martian dust is likely to cause or contribute to many problems during a given mission, but for the landing itself it would fool radar altimeters, which have to measure closing velocity, and would then cause a large plume of material to cover both the landing area and the lander itself at touch-down. Thick dust may also be incapable of supporting the lander. Some parts of Mars have been referred to as 'stealth' regions because of their low radar-reflection properties, and are believed to be regions covered in dust several metres thick. This has been described as 'foo-foo dust,' a Dr Seuss-like term used by Matt Golombek for possible fluffy accumulations of fine iron oxide dust particles that can give rise to drifts of 'red snow'.

Despite all this, the geologists would want to sample exposed rocks, so the ideal site would have scattered boulders from a variety of different sources. If the boulders are too small they will be too badly weathered, and if there are too many of them, the area will be unsafe for a landing. It is important to know where rocks have come from to provide the all-important context, but for a planet of which we know very little, any information is useful, and the ability to date the rocks is critical. It must also be decided why a particular site is suitable, and what work is to be undertaken.

Is it the geological history of Mars, whether life ever existed on Mars, or the history of climate change on Mars? A site that is ideal for geologists may not be suitable in the search for life, and *vice versa*.[12]

**Information is everything**
In the days of Viking, after the landing site had been chosen it had to be certified. The landing site working group would have been hopeful that the site was safe, but could only be certain after additional data was gathered. Sites for the Viking landers in 1976 could only be certified when their parent Viking orbiters arrived in orbit armed with cameras for mapping. But the orbiters' cameras could not detect objects as small as the landers; in fact, the smallest detail that they could detect was 100 metres across, so the risks had to be extrapolated from what could be seen in the images. The images provided context information and the location of large craters, and other useful sources of data, to shed light on surface conditions at the landing sites, came from Earth-based radar installations such as the Goldstone and Arecibo radio telescopes, as well as an infrared mapper (IRTM) on the Viking orbiter. It was early days for analysis of this type of information, and the 'experts' in those fields had a difficult job in persuading the landing site working group to accept the results, often ruling out a particular landing site because the radar return suggested a dangerous surface, even though, in photographs, it appeared to be perfectly safe.[13]

An Earth sample of basaltic AA lava. This texture of lava is broken up into very rough slabs when it solidifies. A landing on this type of material would be dangerous. (Courtesy Lapworth Museum of Geology, University of Birmingham.)

An Earth sample of basaltic pahoehoe lava. The texture of this lava is smooth, similar to molasses. Both AA and pahoehoe are named after the Hawaiian terms for these types of lava. The differences are due to factors such as composition, temperature and fluidity. (Courtesy Lapworth Museum of Geology, University of Birmingham.)

Earth-based radar is limited in the latitudes on Mars that can be studied. Only equatorial zones can be probed, because signals glance off high latitudes rather than bounce back; but at least it offered the promise of information to a resolution of centimetres. The arrival of radar-bearing spacecraft, such as ESA's Mars Express in 2003, has considerably eased the process. Radar provides data on elevations, surface roughness and bearing strengths for landing areas, and will also reveal areas with coverings of dust. Radar can sample thousands of square kilometres at a time, but can also measure average roughness to as fine as 10 cm. Infrared data can reveal surface composition, showing how quickly or slowly the surface materials heat up or cool down. Areas covered in dust cool down relatively quickly after dusk (low thermal inertia), while solid rock will hold the Sun's heat for longer.[12] Colour analysis of the surface can also be used to provide hints about the amount of dust present. Martian dust is red, while the underlying rocks are darker and bluer, so the red–blue ratio measures the amount of dust cover.[14] Laser altimetry – like the MOLA instrument on Mars Global Surveyor – will provide data on surface slopes and surface roughness,[15] and even space telescopes may be used for the gathering of infrared data on landing sites. The Hubble Space Telescope achieved results as good as those obtained by Mars Global Surveyor (orbiting Mars at that time) in 1999.[16]

**Scientific influences on the choice of landing sites**

In addition to all the other considerations and constraints, the different scientific disciplines that would be considered for a Mars mission would also each have their own requirements. For geology and geochemistry, a major type of terrain would be chosen. There are more than a hundred different types of formation on Mars, but this can be narrowed down to eleven materials at primary sites. A single landing could not hope to investigate more than a few of these, but a landing site can be chosen to maximise the numbers of rocks and the topography that can be seen, especially if layering is visible. A landing site near an impact crater or basin would provide the opportunity to sample rocks excavated from the crust, or even the mantle, from a big impact. In 1997 the Mars Pathfinder landing site was chosen for the possibility of the largest variety of different rock types. near the mouth of an outwash channel from ancient catastrophic floods. Geologists termed this a 'grab-bag' site.

Seismology requires multiple measuring stations. These would be closely spaced – say, 100 km apart – for local events, and globally dispersed – 3,000 km apart – to investigate much deeper phenomena. Part of such a network might be deployed on one landing, but a series of landings would probably be necessary. Meteorology would need its own network of measurement stations to study the circulation of the martian atmosphere and the general climate system of Mars. Major dust-storms originate south of the equator, and atmospheric circulation is very different in the northern and southern hemispheres, due to the topography – the two faces of Mars. Both seismological and meteorological measurements would need to be spread over several martian years (global dust-storms, for example, take place only during certain years), and ideally the meteorological stations should be in locally smooth areas.

Exobiology (the search for life) may impose the biggest requirements on a landing site. Water is deemed essential for life, and any landing site should have land-forms associated with water and evidence of sedimentary deposits such as minerals laid down by water. Deposits from ancient thermal springs are high-priority targets due to the chance of finding preserved microbial fossils, and drill cores would be needed to sample subsurface layers to search for fossils or evidence of past life. The sampling of rocks near impact craters also offers the possibility of collecting subsurface material, although the ejecta may be metamorphosed by the shock forces. Life probably required water to exist for a long period of time, so water-carved or deposited terrain, created over a long duration, would be sought; and if the search is for current life then the site would have to have a temperature not far below freezing, the highest traces of water, and the highest atmospheric pressure (lowest elevation). Imaging of the martian surface at a sufficiently high resolution – able to routinely detect objects 30 metres wide and just a few metres wide in high-resolution mode – only became available with the arrival of Mars Global Surveyor in 1997.

The landing site itself would be a circle roughly 20 km wide – the size of a city – for a lander capable of manoeuvring in the atmosphere, and the entire landing site circle would need to meet the criteria for a safe landing. Ideally, this would first require 1-metre resolution imaging of the whole landing area, but this is often not possible. Every time the resolution is halved, the amount of accrued data is quadrupled, so the terrain types in the landing circle are analysed using the highest-resolution images available. Images with a resolution of 1 metre show the very largest boulders when the Sun is low enough in the sky to cast the best shadows, but 10-cm resolution is required to reveal house-brick-sized rocks.[12] The landing circle becomes more elliptical the further from the equator the landing takes place – grazing the poles of Mars rather than plummeting straight into the atmosphere – but any landing site will be chosen to contain interesting features all over the landing circle. Once it has been chosen, the next stage is to reach Mars and, on arrival, deal with any contingencies or alterations to the plan.

## REFERENCES

1  Interview with Alex Ellery, 10 October 2003.
2  *Exploring the Moon and Mars: Choices for the Nation*, OTA-ISC-502, 1991.
3  *Safe on Mars, Precursor Measurements Necessary to Support Human Operations on the Martian Surface*, National Academy Press, 2002.
4  Interview with Michael Duke, 28 January 2004.
5  Carol R. Stoker, *Science Strategy for the Human Exploration of Mars*, AAS 95-493.
6  Morton, *Mapping Mars*, 2002, p. 216.
7  Jakosky and Golombek, First Landing Site Workshop for MER, 2003, 9004.
8  Matt Golombek, personal communication, 2003.
9  Raeburn, *Uncovering the Secrets of the Red Planet*, 1998, p. 129.
10  Kass and Schofield, First Landing Site Workshop for MER, 2003, 9037.
11  Golombek, Jakosky and Mellon, First Landing Site Workshop for MER, 2003, 9017.
12  Interview with Dave Rothery, 6 August 2002.
13  Ezell and Ezell, *On Mars*, NASA SP-4212, p. 320.
14  Ruff and Christensen, First Landing Site Workshop for MER, 2003, 9026.
15  Duxbury and Ivanov, First Landing Site Workshop for MER, 2003, 9025.
16  Morris and Bell, First Landing Site Workshop for MER, 2003, 9039.

# Voyage to Mars

This book is not concerned with a journey to Mars. The method of travel has not yet been chosen, and the hardware has not yet been planned or defined. Mars is a near-neighbour of Earth, and there are times when Mars approaches relatively close to us (though still not as close as Venus); but the problem is that a spacecraft cannot just 'zip' across the intervening gap between the worlds. It might be thought that a launch could be timed to coincide with the close approach of Mars to Earth, and that a rocket could be launched from Earth's surface and aimed at Mars, and then fire its engines all the way until it arrives. But it is not that simple. The distance from Earth to Mars is 56 million km even at such favourable times, and no current rocket can carry enough fuel and oxidiser (propellants) to sustain continuous engine-firing. The planets are also moving around the Sun, and are moving at different speeds. The spacecraft becomes another body orbiting the Sun, and has to go where the Sun's powerful gravity will allow it.

The best analogy is to think of the Solar System as a big deep bowl with the Sun in the centre. Earth is rolling around the inside of the bowl not far from the Sun, but it will not slow down and fall into the Sun because this is a frictionless bowl. A rocket carrying a spacecraft is launched from Earth and has the same speed as our home planet plus a little more. The speed imparted by rocket engines pales into insignificance compared to the speeds of the planets travelling around the Sun. Mars is further away from the Sun than the Earth is, and so the rocket's objective is to climb higher up the bowl while rolling around it. The spacecraft will not slow down and fall towards the Sun in this frictionless bowl, and the spacecraft has all the benefit of leaving the relatively fast-moving planet Earth. But the spacecraft has to gain much more speed so that it can climb higher up the wall of the bowl and move towards Mars. This could be achieved in one manoeuvre after launch, but it would need a very powerful set of engines. It could even be achieved incrementally using a different type of propulsion called ion drive (which is being used by some new robotic spacecraft), but the engines are nowhere near as powerful as traditional rocket engines. Travelling from Earth to Mars still requires that a spacecraft travel up to halfway around the Sun before it raches Mars – a journey of about 400 million km, and not the 56 million km of Earth–Mars close approach.

Humans travelling from Earth to Mars will become more like the heroic explorers of previous centuries. They will know where they are going, they will, in general terms, know what to expect, and they will know how to get there; but this will not lessen the ordeal of the journey. Before we go, a tremendous amount of preparation will need to be carried out: testing the equipment, simulating the potential scenarios, and rigorously training both the flight crew and the support personnel.

## TRAINING

The selection of the crew will be dependent upon a wide range of factors, and will undoubtedly be both politically and financially influenced. Will the crew be international or solely from one nation? Will the command fall to an experienced astronaut or someone from the country that contributes the most money? Even the size of the crew will determine its composition. A large crew will be able to specialise in different areas of the mission, with some cross-training for contingency purposes, but a small crew will have to be more generalist, and able to turn their hand to multi-tasking. It would be unrealistic to speculate on the make-up of the crew, but two factors are certain to play a major role in their selection. Firstly, the crew will need to be able to cope with the isolation of the journey to Mars and be able to work together to ensure their survival and cope with any unforeseen drama both on Mars and back on Earth (see *Living the Dream*, p. 193). Perhaps more importantly, however, there may have to be a trade-off between experience and the individual's career radiation exposure. Sending a team of rookies on such a hazardous journey would be unlikely, but neither could a team of six-mission veterans be sent. While their experience would be invaluable, the radiation dosage that they could potentially face on the mission could prove damaging or even fatal considering their previous exposure.

Without knowing the specifics of the first human mission to Mars, it is impossible to determine exactly what training and preparation would be incorporated before the flight. But we can draw on the experiences of Apollo, preparations for long-duration space station missions, and several Mars training and simulation programmes already underway, to provide us with some idea of the types of training that would be employed. There are numerous training environments and processes that could be employed for human missions to Mars, and it is possible that all of them would be used, or at least evaluated, to provide the first human mission with the maximum chance of success.

### Mars on Earth
Training will include realistic terrain that is analogous to parts of Mars. There are several parts of the world that could be used for this purpose, such as the interior of Australia, Iceland, the Rio Tinto district of Spain (which has high metal content in the ground), and the desert in Utah. For consideration as a training location, the area would have to fit in with what the crew would expect to encounter on Mars, or to at least provide the right circumstances to train for the sciences that would be

The interior of Iceland is remote, barren and uninhabited. It has been compared to the surface of Mars. (Photograph by Andrew Salmon.)

employed, as with the Apollo astronaut geology training in Iceland in the 1960s. This would include the area's appearance (choosing somewhere that looks like the surface of Mars), and somewhere where the geology and/or soil chemistry is similar to that revealed by Viking, Mars Pathfinder–Sojourner, or Mars Exploration Rovers.

The Mars Society is very active with its Mars Analog Research Station (MARS) modules in these types of location. The first was Flashline-MARS in the Canadian high arctic – an impact (meteor crater) site on Devon Island activated in July 2000 – and the second was the Mars Desert Research Station (MDRS) in Utah, activated in February 2002. The third will be the Euro-MARS facility, to be tested in 2004 and then based in the Krafla-Myvatn region of north-east Iceland in 2004–05. The fourth will be MARS-Oz, to be set up at Arkaroola, South Australia, in 2005. In August 1997, NASA–Ames Research Center set up its own Haughton–Mars Project on

The volcanic region near Lake Myvatn in north-east Iceland has terrain similar to this. The Mars Society's Euro-MARS habitat will be located there. (Photograph by Andrew Salmon.)

Devon Island, and was the first to carry out active field science and human factors work with application to human explorers on Mars. Today, the NASA HMP exists alongside the Flashline-MARS facility, and some joint work has been undertaken. NASA also makes use of the MDRS in Utah.

For Mars Society expeditions, a mission briefing pack is prepared at the Mars Society headquarters in Boulder, Colorado, and sent each day by communications link to the crew in the habitation module. The information is received from 'Earth', digested, and talked about by the crew, and the tasks to be carried out during the day are discussed. At the end of the day the crew compare notes concerning their experiences, ready for communication back to 'Earth'.

The Mars Society has also built in a time delay for information sent from 'mission control' in Boulder to the MARS modules. The one-way time delay is 5–20 minutes, simulating the respective positions of Earth and Mars. This equates to a 10–40-minute round-trip between asking a question and receiving a reply, and this can be simulated by relying on e-mails with built-in delays in the 'send' times. The delay times are exactly the same for voice communications and telemetry (engineering or scientific data), and because of these long time delays, the focus in the MARS modules is on independence for humans on Mars. Work is conducted by the occupants of the training facility, who infrequently ask for help from 'Earth'. This may be just once per day and also at the end of an EVA, and it is accepted that there will not be a quick turnaround to receive an answer.

The 'crews' inside the MARS modules are also supported by permanent virtual mission control centres. When the MARS occupants have questions or require advice or help, then the questions are forwarded to a virtual mission support centre rather than to Boulder. They are sent into the ether to an e-mail address, and can be picked up by whoever is on duty in any city in the USA, at any time of day or night. Help is sought to answer the question, and the response is sent back to the MARS module.[1] For real missions to Mars, astronauts on EVA would not be reliant on rapid contact with mission control on Earth, which would simply not be possible due to radio signal time delays. Instead, there would be near-constant contact with the crew of the lander module or the mother-craft in orbit – their local 'virtual mission control'.

**Equipment and procedures**
Apart from on-location training to simulate the surface of Mars, the crew and support personnel for the first human mission will also have to train for the procedures and the equipment to be used on the mission and for the environmental and psychological conditions of both the work on Mars and the voyage there and back. This could include anything from parabolic aircraft flights and simulations of martian gravity to closed environment habitation and isolation tests.

Isolation testing was experienced to some extent by the American astronauts who spent several months as part of the crew onboard the Russian space station Mir in the mid- to late 1990s. It was difficult to arrange regular, reliable link-ups with NASA and family, and it was exacerbated by the cultural isolation of the American in the Russian crew. Long-term studies were also carried out in Russia in the late 1960s – at about the same time as the American Space Task Group studies of missions to Mars as a follow-on to Apollo – and included long-term isolation studies in Moscow and Siberia in facilities such as Bios-2.

But the facilities need not accommodate the crew in complete isolation. In a Closed Environment Life Support System (CELSS), everything that is eaten or excreted is contained within the system. This was achieved to some extent on Mir, and also on the International Space Station where, for example, urine is recycled to produce water to the extent that it can, in theory, be drunk. There has been no attempt at solid-waste recycling, although it could be used in greenhouses, but at present this is disposed of onboard the Progress robot supply craft that burn up in Earth's atmosphere.

**Way stations to Mars**
The International Space Station is also a potential training ground for the journey to Mars and back. It is a high-radiation environment, although not as hazardous as being on the Moon, because the ISS is inside Earth's magnetosphere. Outbursts of material from the Sun and galactic cosmic rays also need to be taken into account, and their study would provide vital data for use in choosing and preparing the crew, as well as the construction of the spacecraft to take them to Mars. A stay of 3–6 months on the ISS would replicate the experiences of Mir crews, and missions to Mars were being considered during the progression from the Soviet Salyut space stations to Mir in the 1970s and 1980s. Cosmonaut Valeri Polyakov's 14 months on Mir (a shorter period than originally planned, due to changed schedules) simulated a Mars return scenario with a one-month stay on the planet. This would not be a very

efficient mission – a long journey each way, and a short stay – but it was quite a common scenario, certainly in Russian plans, and Polyakov's recovery after his mission onboard Mir proved that such a journey would be feasible. The problem with using the ISS is that the training would be in Earth orbit, which would be acceptable for the deep-space journey but of little use for conditions on Mars, unless artificial gravity were to be induced by spinning the training module or its interior.

The problem of zero g could be overcome by training on the Moon if suitable facilities or missions were set up with a future Mars mission in mind. The Moon has low gravity (although the $\frac{1}{6}$ g on the Moon is lower than the $\frac{1}{3}$ g on Mars), and the high solar radiation environment would be good for practice for similar problems on Mars, including the 'duck and cover' procedures needed to ride out solar events such as flares. Equally, wearing a pressure suit and conducting field geology on regolith in a hostile environment relatively close to Earth would be good practice for Mars – particularly in learning to clean and maintain the suit after use. The number of entries and exits from the modules and rovers that would probably be required on Mars or on a lunar simulation of the mission would lead to wear and tear on both the suits and the humans inside, and this rate of wear would be a major factor in planning surface exploration programmes on Mars.

**Life in low gravity**
The effects of low gravity are as high on the study wish-list of mission researchers as that of the human body in weightlessness. This refers to extended operations in $\frac{1}{3}$ Earth gravity (on Mars) or $\frac{1}{6}$ Earth gravity (on the Moon). The first to test it will probably be the Mars Society, with a small satellite placed into low Earth orbit and spun to simulate $\frac{1}{3}$ Earth gravity. There have been many microgravity (zero g) simulations in the NASA KC-135 'Vomit comet', the European Airbus equivalent, and the Russian Il-76, but none in low (non-zero) gravity, and no-one knows what the effect will be of long stays on Mars in $\frac{1}{3}$ g (whether this is a month or one to two years). We know weightlessness is bad for the human body (producing muscle and bone degeneration, for example), but we do not know the effects of $\frac{1}{3}$ g. Is a small amount of gravity sufficient to alleviate bone and muscle degeneration? It could be simulated for up to 30 seconds by the choice of trajectory on parabolic flights, but this would be of only limited use, as such flights would be unable to simulate continuous low-gravity conditions. The use of large water-tanks such as the NASA Sonny Carter Neutral Buoyancy Laboratory for weightless training, combined with a careful choice of weights, might also provide a $\frac{1}{3}$ g simulation, but the drawback is the same as for zero-g simulation: the viscosity of the water. As with all such training, it would be possible to simulate the conditions on Mars only to a limited degree or for a limited time. The only way to really understand the long-term effects of Mars conditions is to go there and experience them.

Low gravity, however, is not the only condition on Mars that would need to be considered. Mars has its own atmosphere, environment and weather conditions, all of which could have potentially hazardous consequences for any human mission. Mars has very cold nights, and the seasons change just as on Earth. Mars-equivalent testing is carried out at the Johnson Space Center or the Ames Research Center, but under laboratory conditions. It may be an outdoor sand-pit, but it is in a 'friendly environment'. Mars is subjected to cosmic radiation and radiation from solar flares, and possibly has dust pools on the ground. The thorough testing of equipment in such conditions would have to take place in extreme-environment simulations on Earth, onboard a space station such as the ISS, or on the Moon.

This leads to the question of just how realistic such training should be. Considering the hostile environment that a mission to Mars would face, the answer would be 'as realistic as possible' – but this would involve some risk. Every minute spent on Mars is close to being dangerous – a severe environmental problem, for example – and although some simulations on Earth could produce the same effect, the ethics must also be considered. The training would be more effective preparation for the potential conditions on Mars, but the elements of risk are high if the training is that realistic. A comparison might be live-fire ammunition training in the military, which is very realistic training but has to be balanced against the potential for accidentally injuring or killing the trainees. Such risky training – such as the 'space tourist' training in Russia – is already in place for space missions. There is no point in sending someone to the ISS without first determining whether they can cope with similar conditions in training, and such candidates wear a pressure suit and undergo tests in a vacuum chamber. By the very nature of the test environment the risk of death or injury is clearly quite high, even under the strictest control and supervision.

**Tools for the job**

An important aspect of a mission is the equipment that the Mars crew would have to use and the procedures they would follow. Given the long communications delay for help from mission control on Earth, the Mars crew would essentially be on their own, so they would have to know that their equipment is reliable and be able to cope with anything untoward or unexpected. The instruments taken to Mars would have to be tested on Earth in 1 g. The devices could also be tested on parabolic flights or on Earth-orbiting craft to simulate weightlessness or martian ($\frac{1}{3}$ Earth) gravity, and the equipment would essentially have to be able to withstand high gravity in launches and aborts, reduced gravity on Mars, and weightlessness on the journey. Such tests could be conducted on the Moon, but the test process would be complicated by the Moon's environment – lower gravity, higher radiation levels, no atmosphere, and no weather.

Equipment for use on Earth will probably function on Mars – something that the Americans in particular would need to consider. The people who design the equipment for use on a US space mission tend, with some exceptions, to start from scratch rather than with something designed for use on Earth to determine whether it will work during a space mission. The result is equipment that costs a small fortune to develop and can be used only in space. The companies that obtain the contract can invest in training, tests and new technologies and produce a very useful end-product, but it is a very expensive and limited product. For example, spanners for use on Earth work perfectly well in space, whereas a spanner for use only in space will be 'tweaked' and may even be improved, but at great expense. On the other hand, a spanner taken to a MARS facility can be used as a hammer as well as for undoing nuts.[1]

Conversely, the amount of equipment taken to Mars must be balanced with the limited space available and the allowable mass, and the design of smaller and more efficient multi-functional equipment may be the key to this balance. Interestingly, however, it was found on both Mir and the ISS that people who are expert at fixing items (and use their garage at home, for example) tend to obtain all the spare parts and salvaged parts from a visiting supply craft and make a repair work bench for the space station. An example is Don Pettit's construction of an Earth-observation camera tracking system to avoid blurring of pictures, so it may simply be a case of assigning the appropriate person in the crew and providing him with a set of tools.

Whatever the equipment, a procedures timeline, including bench tests of the equipment, needs to be developed. How long would it take to deploy an experiment in the field? How long would it take to analyse the samples brought back into the laboratory module? How many samples can be processed before the best are chosen to be brought back to Earth? How much can be achieved in the pressurised area before more EVA is required? How long does it take to suit up or dust down after return from the surface? How long does it take to prepare a meal using the rations and drinks taken to Mars? How long is the planning cycle – receiving information from Earth, analysing it, and preparing to go out to the surface? How much rest time will be available during the day? This type of analysis was carried out in the SMEAT test for Skylab and by the Russians in a 1-g environment, and there were often verbal comparisons during Apollo missions on the Moon, when astronauts commented on the difference between 1-g simulations and the same task on the Moon. The first mission to Mars will glean information on the differences between 1-g training and the same task on Mars, and any later missions will be revised, based on the experiences of the first crew.

Some of this testing of procedures is already being undertaken at the MARS facilities, or will be carried out at new ones. The control of back-contamination of the habitation and laboratory modules will be simulated at the new EuroMars module, with procedures that will review what will be accomplished after EVA, including the handling of samples in the laboratory module. Dust on the samples must be removed, and the EVA suits must be cleaned and thoroughly dusted down (hand-held vacuum cleaners are currently used for this task).[1] For the mission to Mars, the analysis of the samples would also have to be completed before the next EVA so that the results can be incorporated and the EVA timeline or location possibly modified. If sufficient quantities of particular geological specimens have been acquired, the next EVA can be utilised in searching elsewhere for something new.

Sample analysis under sterile conditions will be simulated using what is effectively a small-scale lunar receiving laboratory (like that at the Johnson Space Center). It will be difficult to ensure sterility when on Mars without taking all the technology that is necessary and available on Earth. As with, for example, carrying out surgery in field conditions, a sterile environment cannot be guaranteed.

Finally, operational studies are being carried out at MARS facilities; for example, field geology combined with living and working together in the enclosed environment of the habitation module. There are also simulated emergency procedures tests in the event of a contamination containment breach or a pressurisation leak, or the 'duck and cover' procedures in the event of a dust-storm or solar flare activity. This is concerned as much with group interactions and psychology as it is with equipment tests, and there is a place for testing both the equipment and the people. How would a crew on Mars cope with events on Earth, such as bereavement or a major disaster? Would they be trained to cope with the possibility, or would there be 'counsellors' among the crew-members? This brings us to the human aspect of the Mars mission, and the qualities required for coping with the mission and ensuring its success.

## THE CREW

The Russians are past masters of psychological studies, particularly for spaceflight crews. They are keen to study how crew-members interact together, how the commander interacts with the crew, and how the crew interacts as an entity. It is vital that Mars mission crew-members should be on good terms and be able to work together as a team throughout long periods. This is relatively easy during a short mission of a few days or a week in Earth orbit, but conditions will be totally different during a mission to Mars. It will be the first time that humans will be away from Earth for two years, and it will require a certain type of character to be away from Earth and its comforts for so long. This is where psychological studies and crew compatibility are important, because there will be no view of Earth at all. Being on Mir or the ISS is a day-trip by comparison, because throughout that time the window blinds are open and the Earth can be seen below. Most Earth-orbiting crews spend all their free time staring out of the window, and often do the same while eating meals or exercising. This will not be possible on missions to Mars.

The lessons learned from experiences such as Ernest Shackleton's attempt to reach the South Pole showed how important it was to have a strong leader, and this will also be true of the mission to Mars. Shackleton's mission succeeded – not in the original aims, but simply in that they all returned safe and alive. This was possible due to Shackleton's character.

### Human skills
There is no doubt that human missions are much more expensive than robotic missions when simply compared mission by mission, but this is an incomplete comparison. The training of a human to carry out science on Mars can be utilised for training later crews. The technology adapted by humans to accomplish work on Mars can be used by later crews. It is the acquisition and transfer of knowledge, together with a whole range of other skills and abilities that will make a human mission to Mars more likely to succeed than a purely robotic mission.

Robots can be used on Mars to survey the planet's composition, monitor its weather, and collect samples for return to Earth. This is the role undertaken by the

Spirit and Opportunity MERs and the Mars Sample Return Missions planned for after 2010. Human explorers will still be required for geological field studies or to search for signs of indigenous or fossil life, because humans bring a wealth of knowledge arising from experience and the ability to link a set of unexpected observations. They recognise subtle clues, and can assess and analyse situations; they can integrate input from multiple sensors and see connections; and they respond to new situations and adapt their strategies accordingly. On the other hand they are less predictable than robots. They suffer illness, homesickness, stress arising from confinement, hunger and thirst. In contrast, robots are especially good at repetitive tasks. Like computer programs, they are best at gathering large amounts of data and carrying out simple basic analysis, so they are eminently suitable for reconnaissance. They are also very predictable, and can test hypotheses suggested by the data that they gather. But they are hard to reconfigure for new tasks, and are subject to mechanical and manufacturing errors or mistakes by their human operators.

Human capabilities and human mobility are also important factors. Can spacecraft carrying humans travel faster and further afield than robots controlled from Earth? Robots can achieve a great deal, and can continue working – they do not need to eat, sleep or rest – but if there is a mechanical problem, how can they be repaired? Humans with the appropriate skills can maintain everything in good order.[2]

Humans are infinitely adaptable. When facing problems they have the ability to cope with the unexpected, and can apply thought to solve a problem. Robots, on the other hand, have to carry out tasks following a set pattern, while humans can change the procedure. Humans also have intuition – a psychological trait – a brain skill. People can sense what has to be done without being able to explain why or how they have sensed it. Human adaptability, and the ability to react quickly to changing situations, will be essential for a mission to Mars.

Equally, there will be certain constraints to deal with before the crew is selected. Some crew-members may be rookies, while others will have flown in space before; but one complication is the issue of lifetime radiation dose. The best people to fly to Mars may be those who have never flown beyond low Earth orbit, those who have never flown in space before, or those of a certain age. It is possible that the entire crew could be aged sixty or more – those for whom life expectancy is not likely to be drastically shortened by the increased risk of cancer through radiation exposure. This also brings into play the balance of training and experience. Should the crew be comprised of people who have the knowledge to cope with any given situation, but who might not survive for long after post-flight readaptation due to the prolonged radiation of deep space? Or should it be an inexperienced crew, who will not be subjected to the lifetime radiation dose limit on such a long flight, but who might be a greater liability because they do not have the experience?

**The university of life**
Experience is gained throughout the time spent at college, at university, and then in specialist training for a career. This includes the 'university of life'. It is a certain type of person, in a particular pool, who becomes an astronaut. Those from the military

happen to be very capable when they are trained for other tasks; for example, some of the Apollo pilot astronauts trained in field geology had a natural aptitude for the task. Scientists, on the other hand, have different experience, skills and training. In the Apollo days, those who became scientist-astronauts came from a certain very specialist background. A few had fast jet training and those who did not had to learn, and they were a very eclectic group. Most scientists would not be suitable for work in astronautics, and few of them would even want to train for the job. Some of them are very dismissive about the work carried out during human spaceflight, and regard it as a waste of money. They are keen to have money spent on experiments, and prefer robots to be used for the work. They want only the data, and are not concerned about how it is gathered. They analyse the data, produce papers for scientific journals, and receive the acclaim, the applause and the Nobel prizes. Conversely, there are scientists who would jump at the chance to undertake such pioneering field-work. The scientists in a chosen Mars crew will probably be carrying out active research themselves. The end product of the mission might be scientific papers, possibly written on Mars, and certainly written during the long journey home, to be sent to Earth for publication. Papers will possibly be accepted for publication (and awards) while the author is still in space.

**The value of an engineer**
The first proposals and plans for the Mars Analog Research Stations included crews of only four people – all generalists, and no specialists.[1] This was completely counter to NASA's approach, and NASA was aghast at the idea of there being no specialists. But it has since been discovered, during the field seasons in Canada and Utah, that people's specialities are based on their experience and skills. Those who have engineering skills use them to routinely maintain or repair equipment when necessary. It is an important role for the crews in the habitats: a chief engineer, experienced in construction, plumbing, electronics and electrical repair tasks.

Such adaptability lends itself more to Russian methods than the more rigid American system. The Russians are trained in a certain number of skills, which can be used each day and applied to many tasks. They have a certain number of mental 'tools' which can be used as and when required, and are not trained in only one discipline such as, for example, how to carry out an experiment. The American system has a timeline of procedures for each minute of each hour for the entire day. Each task is assigned a specific period of time, and tasks have to be performed in a certain sequence. When problems occur they disrupt the timeline, and a change of timeline has to be agreed with mission control. On Hubble Space Telescope servicing work, Mir and the ISS, the benefits of the Russian approach have been appreciated, and for missions to Mars this flexibility may be vital, especially if the chosen final crew is small. The original MARS concept of four-person crews has now been increased to six, but whatever the complement of crew-members, they will essentially work as an autonomous, self-guided unit, able to make decisions *in situ* without having to wait for orders from Earth. Astronauts have the ability to do this. They are independent people, but they are disciplined to follow orders and obey the rules. There will need to be a commander – a single point of contact to settle disputes – and there will be scientists and an engineer (the chief engineer). No serious work has so far been undertaken on field medicine, but such training will take place. Otherwise, the crew will probably split the tasks as needed, being adaptable and flexible enough to complete whatever tasks are necessary for any particular process.

The assigned crew will also be responsible for much of the planning. During the Apollo days, the Principal Investigators – the scientists with specific work to carry out – received funding from their national governments, while the space agency received funds for the mission. The samples were acquired, brought back to Earth, and sent to the lunar receiving laboratory, where they were cut up as smaller samples and sent to various institutions around the world for analysis. Those who did not originally request samples could apply for funding to carry out sample analysis, and funds are still available for the analysis of Moon rocks collected between 1969 and 1972. In contrast, the scientists analysing samples on a mission to Mars might actually be the astronauts on the mission, or they may be dependent on trained astronauts who can select the types of sample required while in the field, or who can study samples in the lander laboratory and conduct EVA to collect better ones. Sample selection would be based on personal choice and basic training in biology and geology. The Mars Society field-trips in the Arctic or Utah include crew-members who are given general instructions about what is required, such as a search for rocks that have lichens on them or which are of a certain colour. But the power of intuition again comes into operation. They 'feel' that a certain rock is worth examining – an instinct that they will find what they want in a certain location. These are hidden signs that are picked up by the eye and brain.

In this media-hungry age, every member of the crew will need to be comfortable with public affairs work. Diaries and debriefing notes, and possibly even autobiographies, would be written on the return flight. Once the crew returns to Earth, there is also the question of life after Mars. They would have to cope with endless public appearances, and so other skills would be related to public speaking – another aspect completely different from piloting skills or scientific knowledge.

One thing is certain. Whatever the crew composition, their training will last for several years. There are so many factors to take into consideration that it will take that long to organise them, incorporate them in training, and ensure that all crew-members work well together to cope with any situation that might arise. Once they are on their way to Mars, there will certainly be no turning back. The whole mission will stand or fall by the compatibility of the crew and the quality of their training.

## THE JOURNEY TO MARS

Once the decision is made to go to Mars, it will still be several years before the mission is undertaken. The physical alignment of the planets will dictate when the launch windows are available, and the training of the crew must be thorough. The design of the spacecraft may well be updated as new technology comes on line, or the decision might be made to go with what is known to work rather than run the risk of new equipment not functioning properly. A major influence in the design and construction of the spacecraft will be the method used to reach Mars.

The orbits of Earth and Mars around the Sun are not circular. They are elliptical – the orbit of Mars much more so than that of Earth (This discovery, by Johannes Kepler in the early seventeenth century, marked a watershed in astronomy.) The orbit of Mars is also tilted at 1°.85 with respect to Earth's orbit (the plane of the ecliptic), and a spacecraft travelling in interplanetary space is also orbiting the Sun. So, a journey from Earth to Mars is not a trivial matter.

**Crossing the void**
The most propellant-efficient trajectory is called a Hohmann transfer orbit (after a theory proposed in 1925 by the German engineer and mathematician Walter Hohmann). It is also known as a minimum energy transfer orbit. A pulse of rocket power near Earth sends a craft on a long, slow cruise through interplanetary space, in an elliptical transfer orbit that just grazes the orbits of Earth and Mars. The launch is timed so that the spacecraft reaches a point in space halfway across the Solar System on the far side of the Sun, just as Mars reaches the same position. These launch opportunities occur once every 26 months. When the craft arrives at Mars, another engine pulse enables it to enter orbit, although it could, of course, sail straight past and fly around Mars before eventually returning to Earth. This free

The long flight to and from Mars will be based upon experiences from long-duration space station missions. Cosmonaut Valeri Polyakov currently holds the individual space flight record of 14 months – more than the expected duration of a trip to Mars.

return trajectory is the type of path followed by the early Apollo flights to the Moon. In the event of a problem, the craft will eventually return to Earth without another engine-firing until required to enter Earth orbit or slow down for re-entry.

An interplanetary spacecraft is being acted upon by the all-encompassing gravitational field of the Sun and by the gravitational fields of the other planets. To simplify navigation, the 'patched conic method' could be used. A spacecraft with its own propulsion is said to follow a trajectory rather than an orbit. (This is a point of semantics, because a spacecraft's propulsion is mostly only used at the origin and destination points.) The patched conic method divides the trajectory into 'legs', dealing with just one planetary body at a time. Each planet has a 'sphere of influence', within which the planet is the dominant gravitational force. Earth's sphere of influence is 0.93 million km in radius – the first leg of the journey – outside which the Sun is the dominant force and controls the second and longest leg. The gravitational influence of Mars is 0.58 million km in radius, and this is the third and final leg of the journey.

Trajectories to Mars other than Hohmann types are called 'fast conjunction' and 'opposition'. These could be described as 'Solar System snooker' – moving from

A nuclear-ion engine proposed by Korolev's OKB-1 design bureau for use on a Soviet manned mission to Mars. (Model at TsNII Mash, Korolev; photograph by Andrew Salmon.)

planet to planet like a snooker ball on a table. Planets are said to be in conjunction or at opposition depending on their position relative to the Sun and the Earth. Mars is at opposition when the line-up is Sun–Earth–Mars, and in conjunction when it is on the far side of the Sun and the line-up is Mars–Sun–Earth. (Neither Mercury nor Venus can be at opposition, because their orbits lie between Earth's orbit and the Sun). At conjunction, Mars lies on the opposite side of the Sun at its furthest point from Earth, and cannot be seen; and close to conjunction it is very difficult to see. This would also interfere with radio communications. The types of orbit faster than the Hohmann orbit – open-ended hyperbolic trajectories – provide faster transfers between worlds, and require a much higher launch velocity than a Hohmann orbit.

Opposition-type trajectories certainly reduce total mission length, although up to two years is spent in interplanetary space, which is not advisable at times of high solar activity. A peak of solar activity does not necessarily coincide with solar maximum, and can occur some time before or after. (For example, there were several flares and coronal mass ejections in October–December 2003, although solar maximum occurred in 2001.)

Fast conjunction trajectories minimise the length of time spent in interplanetary space, and this method – both inbound and outbound – was chosen for both the Mars Direct scenario of Robert Zubrin and the Mars Society and for the NASA Design Reference Mission.

**Celestial mechanics**

The timing of launch from Earth depends crucially on the positions of Earth and Mars in their orbits, and is calculated using celestial mechanics. This is combined with the type of spacecraft engine and the quantity of onboard propellants available for firing of the departure engine. The craft will then be on 'the road to Mars'.

To give a craft more momentum, gravity assist techniques can be used. The craft flies relatively close to a planet to utilise the the planet's gravity and orbital momentum. If the planet was not orbiting the Sun, the speed increase gained by falling into the planet's gravity well would exactly match the decrease due to the climb out on the other side, so there would be no change in velocity. But the planet *is* orbiting the Sun, and the spacecraft therefore 'robs' a little of the planet's orbital momentum and it is 'tossed' onward by the planet, which falls towards the Sun by an infinitesimal amount. Depending on whether the craft approaches the leading or trailing side of the planet, it can either shed or boost its velocity, and the planet's orbital momentum increases or decreases accordingly. To move on to Mars, the technique is to approach Venus (within Earth's orbit) and be 'tossed' further out from the Sun. Gravity assist can also be used to change both the speed of a spacecraft and its trajectory. For analogy, consider a ball being tossed from the roadside to a passenger sitting in a moving convertible car with the top open. The ball is then caught and thrown to someone else standing on the roadside, so the speed of the ball is now its speed when it was thrown from the car plus the speed of the car. The original direction of the ball has also changed.

Much of the above relates to free-flight trajectories (unpowered during flight) and, for Hohmann transfers, to minimum energy trajectories. The use of technologies such as low-thrust, continuous-power engines will provide many other options when perfected. Traditional chemical rockets travel through the Solar System using (not surprisingly) chemicals in a rocket engine. An oxidiser and a fuel are ignited to burn, and an impulse is produced using a nozzle.

A mission could start from the surface of the Earth, or it could start from Earth orbit after being assembled there or delivered. Some consider that a start from Earth orbit is much more efficient, while starting from Earth's surface is preferred by others for simplicity. Robert Zubrin's Mars Direct plans allow for a direct flight from Earth's surface to Mars, using very large launch vehicles comparable with the Saturn V used for sending the Apollo astronauts to the Moon.

Yet another possibility is assembly of the Mars expedition at one of the Earth–Moon Lagrangian points (L1, E, M). These are gravitationally stable points where a craft can remain with minimal disruption from other bodies in space. Finally, there is the spaceport in a 'cycler-orbit' popularised by former astronaut Buzz Aldrin. A spaceport is set endlessly travelling between Earth and Mars, and ferries carrying a crew, supplies, equipment and other payloads, join it at either Earth or Mars. This would be a more mature infrastructure, and would probably not be used for the first human mission.

The time spent from Trans-Mars Injection (TMI) to Mars Orbit Insertion (MOI) is variable, as it is dependent on the positions of Earth and Mars at launch and the type of propulsion used to cross the gulf of space between the two planets. A propellant might be used for departure from Earth and not used again until arrival at Mars, but ion drive has the advantage that very little propellant is required for an engine-firing, so that it could even be fired continuously from Earth to Mars. A solar sail carries no propellant, but instead uses the pressure of sunlight. However, its disadvantage is that it takes a considerable time to increase and decrease its speed.

**Keeping the kids amused**
The flight from Earth to Mars – possibly via Venus and approximately eight months long – does not have to be a time of complete inactivity. To keep costs down, communications sessions with the craft may be kept to a minimum, and to further constrain costs there may be no scientific instrumentation for use during the cruise phase. The minimised need for constant real-time communications through the Deep Space Network further reduces costs.

Other factors to be considered are psychological. The crew should be kept active and productive, and safety needs may demand constant monitoring of telemetry. The incremental cost of conducting science during the cruise and the Venus fly-by may be small enough to justify it for psychological purposes.

Cruise science is difficult to justify when considering the total cost of the mission, but it seems sensible if the craft has sufficient launch mass, stowage volume, tracking time and data acquisition facilities. Cruise science carried out on deep space missions includes the following:

*Gamma-ray bursts* Pinpointing the position of these objects in the sky by means of triangulation using detectors in several different locations in the Solar System. A longer baseline for joint observations allows more accurate measurement of the GRB locations.

*Gravitational wave detectors* Sensing the passage of a gravitational wave (as predicted by Einstein) using the radio transmitter of the craft and the Deep Space Network. In the future there will be ultrasensitive, dedicated gravitational wave detectors such as the three laser interferometry satellites (LISA). Such spacecraft will be much more sensitive than crewed spacecraft because there would be no interference from crew support systems or disturbance from the crew's movements.

*Solar observatory* The Soviet-led Phobos 1 mission carried out studies with a solar telescope. Such instrumentation is an advantage, as it can be used to detect ejected solar material heading towards the craft. On a future mission to Mars, visible light, X-ray and ultraviolet telescopes with spectrometers would be supplemented by radio and charged particle monitors and radiation dosimeters.[3] The minimum requirement on a Mars craft would be a solar X-ray imaging telescope for solar flare alerts,[14] but a key point is that the monitoring should be automated so as not to overburden the crew with repetitive tasks. Onboard data analysis and interpretation is important, due to the radio delay times involved. The output from the nuclear furnace that is the Sun is a fairly constant output background of both light and heat, but the Sun is an active star, with a cycle of activity – maximum to minimum to maximum – of approximately 11 years (although the magnetic field changes polarity at the end of each cycle, and it may therefore be considered to be a 22-year cycle). Some of the most active constituents of the Sun are sunspots, solar flares and coronal mass ejections (CME). A flare is a sudden release of energy that ejects material from the Sun, while a CME consists of a huge quantity of material erupted from the Sun's corona as a gigantic 'bubble' of ionised gas (plasma). Flares and CMEs produce particles and radiation that can affect astronauts, communications and equipment. Solar activity would be closely monitored by observatories on Earth, other craft in solar orbit, and the Mars-bound craft itself. Solar particle events (SPE) are the main cause of the early effects of space radiation on humans. A spacecraft can be shielded, but the material used for the shielding can cause fragmentation of the particles and consequent increased exposure, so the craft will also need a 'storm shelter' for the crew. Racks of food, water tanks, consumables and waste products could also be used as shielding.[3] Most of the energetic particles arising from a flare are α-particles and ß-particles, which can be stopped by a few centimetres of shielding, and the craft can also be turned in the appropriate direction.

*Mars meteorological observatory* The Mars-bound craft could carry out detailed weather observations of the planet as it approached, to record, for example, the genesis of a dust-storms. Weather satellites would probably be in place at Mars before a human mission set out, but the participation of the crew in such observations would be of psychological benefit.

*Earth observations* Another psychological boost would be telescopic study of the home planet by the crew. Earth would soon become noticeably smaller, but even an image a few pixels wide on a monitor would provide another connection with the rest of humanity as the crew listened to the news being fed to them.

*Dust-stream detection* The Ulysses spaceprobe detected a stream of dust entering the Solar System from interstellar space. Both the Ulysses and Galileo spaceprobes monitored another stream of dust originating from Jupiter's moon Io.

*Particle and field measurements* Space is not a perfect vacuum. Ions are accompanied by radiation, and there is even an interplanetary magnetic field. Solar flares and coronal mass ejections throw material out fast enough for it to escape the Sun's gravity. Cosmic rays fill the vacuum of interplanetary space, but there are fewer cosmic rays at times of high solar activity, and *vice versa*. High-atomic-number and high-energy (HZE) particles pose a risk of damage to the human central nervous system,[5] and cosmic rays, consisting of heavy and comparatively slow-moving nuclei, arrive incessantly from all directions. They are more destructive than α-particles and ß-particles, and protection against them requires shielding several metres thick – which is not practical. The dose may lead to a 1% increase in the risk of succumbing to cancer later in life (this is already 20% for non-smokers), but the risk from cosmic rays is not well understood at present.[6]

Engines are fired for mid-course corrections while travelling between Earth and Mars. This is to correct (not far from Earth) for errors in the Trans-Mars Injection transfer orbit insertion engine firing at Earth, and for gravitational effects between Earth and Mars. The latter are probably known in advance, but are monitored by the Deep Space Network.

**Medical implications**

The medical problems on a long flight to Mars depend partly on whether there is artificial gravity on the spacecraft, such as the centrifuge on the *Discovery* in Arthur C. Clarke's book (and film) *2001: A Space Odyssey*. Other proposals include Robert Zubrin's design for Mars Direct – splitting the craft into two parts joined with one or more tethers and then spun. In this case, too small a centrifuge or spacecraft spin radius produces unwelcome coriolis forces, and the crew would be violently sick as the rotating environment led to nausea, fatigue, and mood and sleep disturbances. A microgravity craft provides more habitable volume (floors and ceilings are available for use), and artificial gravity provides Earth-like conditions.[4] But it is possible that humans will not require gravity for 24 hours per day in order to remain healthy. An onboard short-arm centrifuge may be all that is required.

The air and water systems will be closed cycle, but the solid-food cycle will be open. Space station studies involved 2.1 kg of solid food per day per crew-member, and 4.5 kg per person per day with allowance for system losses, unrecycled consumables and fixed system mass.[7]

If there is no artificial gravity then the crew will need to undergo 2–4 hours of physical exercise each day to minimise bone and muscle wastage, just as for space station crews. If an astronaut is injured and unable to exercise during this time, this could lead to severe deconditioning and the possibility that he or she may not be able to land on Mars. The g-forces upon arrival at Mars and during landing will test the health of the crew and the effectiveness of their conditioning. In the event of an abort, after which the crew will not land on Mars but will instead fly around the planet and return to Earth, they would be subjected to as much as three years of weightlessness.

The medical consequences of a long-duration microgravity spaceflight include decreased muscle strength, cardiovascular deconditioning, increased potential for kidney stones, bone loss, body fluids shifting to the head, decreased fluid and electrolyte levels, decreased red blood-cell count, neurological effects and immune system changes.[8] The debilitating effects of any or all of these may need to be corrected by long-term rehabilitation on return to Earth, or, in the extreme, may even result in a one-way trip for the astronauts, who would not be able to return to Earth at all.

**Mind over matter**

A mission to Mars will be the first time that people travel so far away that they will not be able to see details on the surface of the Earth. From low Earth orbit, the Earth is always very close, and even from the Moon it still appears quite large – it can be occulted by a thumb held at arm's length. The clouds, oceans and continents are visible. and it can be seen turning over several hours. Upon leaving Earth orbit, and with the Earth shrinking in size, it will be a completely new and different experience. The Earth would appear smaller, and Mars would be just a bright object with a reddish tint. This physical and psychological remoteness would be combined with the knowledge of being in a dangerous environment.

Conversation would soon become strained due to the time delay between a question being asked of the crew and their reply being received on Earth. Conversations may become e-mail-based – disjointed but manageable – but communications are far easier if those involved can see each other. News from home might be received only once a week or every few weeks, and even the sounds of Earth – wind, waves and birdsong, for example – would be missed. A comparable psychological experience may be that of submarine crews, who have no sense of the

Mars expedition living quarters. The main cabin module for a 1960s OKB-1 manned Mars mission design. (Model at TsNII Mash, Korolev; photograph by Andrew Salmon.)

time of day or night or time of year. However, they eventually experience life in the open air again, whereas those on a journey to Mars would be faced with an alien environment when they arrived, and no-one waiting to greet them.

Once the engines have been fired to leave Earth orbit, there are few abort options. It would require too much energy to turn around and return to Earth, so the best that can be hoped for is to continue to Mars, swing around the planet, and begin the long journey back to Earth. It is just conceivable that there could be an 'abort to the surface', in which the crew would abandon ship, land on the surface of Mars, survive on local resources and whatever provisions they have taken with them, and wait for a rescue mission.

### Radiation

The hazards to humans on long journeys in space include radiation.[5] Outside Earth's atmosphere, and especially beyond our protective magnetic 'bubble', there are two main hazards: transient solar particle events (SPE) (mostly protons from solar flares and coronal mass ejections), and the continuous high-energy but low-level background of galactic cosmic radiation (GCR). Amidst the GCR are HZE particles, which have both high energy and high atomic number (Z), and are therefore doubly dangerous. They are, however, few in number and can cause only local damage within the body, but they pose the greatest cancer risk and can cause damage to the central nervous system. GCR has the greatest potential for penetrating through radiation shielding. Solar particle events are at their most intense when the Sun is more active, towards and at solar maximum.

The high-intensity radiation from a solar flare kills cells immediately. This may be within a few hours due to acute radiation sickness and effects on the central nervous system, or a few days due to loss of white blood-cells or gut wall-lining cells. Cosmic radiation is actually pushed away by strong solar winds, so levels decrease at solar maximum; but on balance, the times of solar maxima are the worst for human space travel, and should be avoided. During a solar proton event, astronautical activity may be limited for a time lasting from several hours to a few days – including the need to seek shelter from it while on the surface of Mars! The largest emissions are coronal mass ejections, when huge amounts of material are ejected from the Sun. Countermeasures include shielding and a storm shelter on the spacecraft, and a 'solar patrol' like that employed during the Apollo missions to the Moon and on ESA's SOHO satellite that has pioneered the study of active regions on the Sun. A network of satellites in orbit around the Sun – formerly believed to be essential – is therefore not necessary.[4] The shielding on the spacecraft or lander must not give rise to significant secondary radiation, such as neutrons, inside a vehicle or habitat, because it is significantly damaging to living tissue in its own right, and multilayered neutron-absorbing materials would be required. Radiation protection would also be provided by propellant and water tanks on the Mars-bound craft, and protection against GCR could be provided by storm shelters which could be occupied by the crew on a rotating basis during transit.

The effects of the radiation include cancers, changes to the central nervous system, formation of cataracts, genetic (DNA) effects that can be inherited, and detrimental effects on other organs. High doses of radiation can give rise to early effects such as nausea, but effects from lower doses, such as cancers, may not be apparent until later in life. In 1967, NASA's Space Radiation Study Panel published a report[9] which in 1970 was used by the National Research Council and NASA to set recommended space radiation exposure limits,[10] including a whole-body career limit, skin limits, and eye-lens limits. Then, in 1989 the National Council on Radiation Protection and Measurement issued a report providing guidance on career limits for use by NASA.[11] Operations in low Earth orbit are subjected to trapped radiation, but deep space operations are primarily influenced by cosmic rays.

### EVA on the way
There will probably be no planned EVAs during the flight to Mars, but unplanned (contingency) EVAs are possible; for example, because of damage by micrometeorites. However, radiation and particles from solar flares and coronal mass ejections would be a danger to astronauts outside the spacecraft, and warning times would be as short as 30 minutes.[4]

## ARRIVAL AT MARS

Any part of the spacecraft that remains in Mars orbit will be placed into a parking orbit around Mars. For example, this might be the Earth return vehicle and interplanetary propulsion unit (such as a nuclear reactor). Modules such as the surface science laboratory or habitation module would possibly be delivered to Mars in advance of the astronaut-carrying vessel. These would land on Mars directly upon arrival, without the need to enter orbit, and would be checked by robots before the main craft arrived at Mars. Some mission scenarios have the crew landing directly upon arrival at Mars, with no crew-members left in orbit, which would minimise the radiation dosage.

### Approach speed
The mission would arrive at Mars at great speed, but excess velocity can be shed by several means. One option is for the craft to be rotated and the main engine fired in an extended burn in the direction of travel. This is the most wasteful option, because a large amount of propellant would need to be hauled on the journey, to be used only upon arrival.

The craft could also use aerobraking, by dipping into Mars' upper atmosphere for a short time during each orbit, after a short burn of the main engine places it in a high elliptical orbit. Over many orbits (and many months) the apoapsis of the orbit would be lowered. and the orbit circularised to the degree desired. This technique was used for robotic spacecraft such as Mars Global Surveyor and Mars Odyssey, but careful monitoring of the global weather on Mars would be required so that aerobraking could be fine-tuned. Drag forces would be created using the structures and protuberances (such as the solar panels) of the mother-craft, but this would require care in not subjecting them to excessive stress. Each pass through the upper

Aerobraking design for a Mars shuttlecraft to travel from the interplanetary craft into the martian atmosphere for a surface landing. (Model at TsNII Mash, Korolev; photograph by Andrew Salmon.)

atmosphere would need to be sufficiently gentle in order to obviate any need for a protective shield for the main craft.

Aerocapture is an extreme manoeuvre by which a craft arriving in an hyperbolic orbit can be moved into an elliptical or circular orbit. This manoeuvre is similar to aerobraking, but is accomplished over just one pass through the upper atmosphere, at an altitude of only 50 km above the surface of Mars. As with aerobraking, speed is lost by friction, but in this case the atmospheric entry corridor is narrow, like that of a spacecraft re-entering Earth's atmosphere. Entry at an angle of more than 1° from the correct path would be fatal, as the craft would either burn up due to frictional heating (if too steep), or bounce off the atmosphere into space (if too shallow). Knowledge of martian weather conditions upon arrival would be critical, and an aerocapture shield structure would be required on the mother-craft to generate drag and to shield the rest of the structures. To raise the periapsis outside the martian atmosphere, the engines would be ignited after orbital entry, and the craft would remain in an elliptical parking orbit.[12]

### Heavy pressure

Such braking manoeuvres would place high g-forces (possibly 5–6 g for up to an hour) on the crew in the mother-craft. If the transit from Earth has been under microgravity conditions over several months, this will be a tense time. Bone and muscle degeneration is a consequence of long-duration spaceflight in microgravity, and exercise and diet can ease the problem only to a small degree. By comparison, Earth re-entry for Space Shuttle crews is limited to 1.5 g for 20 minutes, while Soyuz capsule re-entries subject the crew to 3.8 g.[6]

The crew will need to don full pressure suits ready for the main engine burn, and main engine check-out will have to be extensive. After the long journey to Mars, and prior to the approach, it might also be necessary to carry out EVA inspections of the main engine systems to check for damage by micrometeorites. The braking rocket engine burn would probably take place over the night side of the planet, and the crew will have stowed all loose items and equipment in advance of the burn to prevent injury from flying objects during deceleration.

By this time the crew will have to act upon their own intitiative and without immediate consultation with mission control, because of the extended delay in communications. If the main engines fail, then the craft will have to continue around Mars in a free return trajectory, to return to the vicinity of Earth eight months later.

## ORBITAL OPERATIONS

The crew of the mother-craft would be exposed to cosmic rays and the full effects of material ejected from solar flares and coronal mass ejections. The time spent in Mars orbit would be added to the journey time to and from Mars, and during the extra time (possibly eighteen months) spent waiting for the next launch opportunity for the return to Earth the crew would be subjected to a significant dose of radiation.

### Science from Mars orbit
The crew in orbit could also serve as a mission control for landing parties and for teleoperated robots, and would act as the communications relay from Mars to Earth. The commander of the orbiting vehicle could act as command authority for the landing party, in which case the status of the lander would be continuously monitored. Maximum autonomy of the astronauts on Mars would be aided by the development, over the next twenty years, of systems that predict most of their problems before they arise.[13] It is just possible that small communications satellites in Mars orbit – like the Tracking and Data Relay Satellite (TDRS) network around Earth – would offer almost round-the-clock contact with those on the ground. The Mars-orbiting satellites could be released from the mother-craft, and would be of use whether or not there is a mission control at Mars.

Robots teleoperated by astronauts on an orbiting spacecraft would obviate fears of forward and back contamination. Humans and their spacesuits might contaminate the surface of Mars with microbes and biochemicals (and considerably hinder the search for indigenous life), but robots would not be infected by martian life while working with samples on the surface.[14] It might seem an anathema to travel an enormous distance to only orbit Mars and not set foot on the surface – particularly for the first human mission – but orbital studies may be the only science permissible during human missions, due to possible contamination of surface materials by microbes brought unintentionally from Earth. While a high level of sterilisation is possible for robots – the Viking landers, for example – it is certainly not practicable for human missions.

Robotic probes – particularly Viking 1, Viking 2, Mars Global Surveyor and

One of many designs of a spacecraft in Mars orbit supporting surface activities by robots and humans. (Courtesy Pat Rawlings.)

Mars Odyssey – have already studied the planet extensively, but the Mars mission mother-craft could usefully supplement the landing parties. The type of instruments carried could include those that maximise the use of a large spacecraft – those that can make use of very high-bandwidth communications and have very high-power energy requirements. The Apollo SIMBAY made use of astronauts on EVA to retrieve film and data canisters, but this would be unnecessary with high-bandwidth digital devices. Analogous of this is the Jupiter Icy Moons Orbiter (JIMO) mission under project Prometheus, on which a nuclear fission reactor provides enough power for large synthetic aperture radar and a high-bandwidth laser communications system sends the digital data to Earth.

Spacecraft in Mars orbit during 1988–2005 – Phobos 1, Phobos 2, Mars Observer, Mars-96, Mars Global Surveyor, Mars Climate Orbiter, Mars Odyssey, Mars Express and Mars Reconnaissance Orbiter – carried instruments including a visible-light camera, a mineralogical mapping spectrometer, ultraviolet and infrared spectrometers, thermal imaging cameras, gamma-ray spectrometers, neutron spectrometers, radar (including subsurface sounding radar), magnetometers, plasma-wave analysers, charged-particle spectrometers, lasers, radar altimeters, and dust counters.

## Work by astronauts in orbit

The precedent for work by astronauts in orbit around Mars is the J-class Apollo missions (Apollo 15, 16 and 17) using the Command Module Pilot in lunar orbit. Similarly, Earth orbital science was carried out by astronauts on Skylab, the Salyut space stations, Mir, and the Space Shuttle.[15] Significant environmental and Earth-observation studies utilising the International Space Station (other than a continuation of the Shuttle and Mir photographic work) have yet to materialise, and the science carried out on the ISS has primarily involved biology, physiology and materials.

The advantage of having astronauts rather than robots carry out the work is that the human eye and brain are far superior to their digital counterparts. Moreover, a much greater bandwidth is available for communications, and more electrical power can be provided for use by scientific instruments.

The main focus of Mars orbital science will be remote sensing (probably utilising some or all of the instruments mentioned above) and *in situ* environmental sensing. Some of the work that could possibly be undertaken includes monitoring of the Sun, and the weather on Mars, for which the orbital spacecraft may also serve as a communications relay. Solar storm warnings from Mars orbit will provide little extra advance notice of threatening conditions for the crew on the surface; but they might be so busy with their work that they do not have time to monitor solar activity, and could even be enshrouded by a dust-storm. Automated radiation alarms may negate the necessity for human monitoring, and a network of satellites in solar orbit would provide additional time for those on the surface to take shelter. During the Apollo days, this role was undertaken by Pioneers 6–9 in solar orbit, and storm warnings were passed to mission control in Houston. Martian dust-storm warnings would be provided by the equivalent of Earth's weather satellites, and could be supplemented by visual observations by the astronauts on a craft in Mars orbit; but the task might be automated by pattern-recognition software using images sent from Mars. The delay time incurred in sending data from Mars to Earth might be tolerable, but this could be another situation in which a delay of just minutes would make a critical difference.

Mars orbital science, conducted by humans, would be a bonus. It could easily be carried out by robots, but as the craft is there it could be fitted out for remote sensing or *in situ* space environment measurements; and the data may even place other surface measurements in context. Ground truth from those on the surface would be compared with the view from orbit in much the same way as Earth observers on the ground, ships at sea, and aircraft and helicopters in the air, were used for comparison with Space Shuttle or Mir observations. Images obtained at one location can be matched with a synoptic or global view.

*In situ* measurements would be taken of (operational) radiation levels – including cosmic rays, solar particles, and trapped radiation (although Mars has no radiation belts comparable to Earth's Van Allen belts) – magnetic fields, and electrical fields. The orbital craft may even attempt to create and investigate artificial aurorae by stimulating the environment with plasma or electrons and charging up the craft while passing through the martian environment. Remote sensing would include observations of the Sun at ultraviolet and X-ray wavelengths, the night-time air-glow above Mars, the total spectral irradiance at Mars (which should be similar to that near the Earth), atmospheric chemistry in the lower, middle and upper levels by viewing the Sun at sunrise and sunset on each orbit, radar mapping, and meteor activity in the atmosphere.

**Telerobotics**

Humans can operate robots by telepresence – independently, or with joysticks. Teleoperation relies on a short time-delay (time of flight) for signals to travel between the operator and the robot. This is impossible from Earth to Mars, but may be possible from Mars orbit, from Phobos to the surface of Mars, or from a lander or base on Mars to rovers on other parts of the planet.

Rovers at locations thousands of kilometres apart could be teleoperated, for example, on the summit of Olympus Mons or in the layered deposits around the northern or southern polar cap. Teleoperation would permit greatly expanded

Marstrain, a 1960 OKB-1 proposal for a surface 'train' of vehicles on a one-year exploration of Mars. Such vehicles could be occupied by the human crew or controlled from orbit if conditions were not deemed safe enough for the crew to land. (Model at TsNII Mash, Korolev; photograph by Andrew Salmon.)

mobility for a human expedition. After landing on Mars, the operators will not even need to leave the landing craft, and if conditions are deemed unsuitable for sending the crew to the surface, some exploration could still be conducted from Mars orbit. This may also be vital after a landing if the crew has spent the journey to Mars in microgravity, with the robots being teleoperated while they reaccustom themselves to gravity. Real-time control of rovers would be used for delicate operations or when terrain becomes more difficult, but otherwise they could be monitored and left to their own devices based upon the day's programming. Teleoperation demands high-fidelity, high-bandwidth communication links for the virtual-reality operator immersion displays, but would allow exploration that would be considered too hazardous for humans – such as scaling the canyon walls of Valles Marineris to search for sedimentary rock layers.

Teleoperated robots would be autonomous except during critical periods of operation, when they would be controlled by a human. This is a division of labour, utilising the best capabilities of robots and humans. The robots would carry out the difficult and dangerous work, and the humans would be the supervisors and would make the choices.[1] A robot could be given general instructions, programmed to indicate difficulties, request more information, and provide reports. Humans would then assess this information and decide on the next or best course of action.

Total immersion of a human in a virtual-reality environment, using data gloves and visors, is also a possibility.[16] Again, however, such a system would not work when there is a time-delay in communications or when very high-bandwidth communications are not available. Immediate communication from Earth to Earth orbit is practical, and to the Moon from Earth is quite difficult; but immediate communication from Earth to Mars is impossible. Vision would work, but feeling what the robot is sensing would not.[2] Ideally, humans on a craft in Mars orbit, on Phobos, or on the surface of Mars, would control the robots.

Teleoperation is not simple – even on Earth. The two Soviet Lunokhod rovers were controlled by a team of operators via a radio link, which was possible because of the short time-delay of signals to and from the Moon. But it was very laborious, and the operators (six or eight of them working at the same time) became very tired. It seems that driving the Lunokhod with a time-delay was completely counter-intuitive, and better results were obtained by those who were not automobile drivers – which was the primary qualification. The half-second delay was tolerable, but teleoperation is exhausting because there is so much to do in controlling the movements of the vehicle, the cameras and the sensors. On a mission to Mars it may be deferred until later in the mission, depending on the condition of the crew at arrival.

In contrast, the Sojourner rover was given instructions by controllers at the Jet Propulsion Laboratory. The camera that provided the images which were used by the controllers was mounted on the immobile Mars Pathfinder lander, which delivered the rover to Mars. Sojourner's movements were therefore limited by the camera's field of view. The 2003 Mars Exploration Rovers had cameras onboard, but still had to rely on images being sent back to Earth so that the controllers could plan the rovers' movements from place to place. They had limited onboard intelligence for deciding how to proceed to avoid hazards – and current robots take a very long time to move around. The Spirit and Opportunity rovers on Mars, for example, explored an area of just a few square kilometres in a little over six months – but an Apollo astronaut in a lunar rover could cover the same area in one day.

### To land, or not to land
After enduring the long journey from Earth to Mars, why would the astronauts want to only orbit Mars rather than land on the planet? The Apollo 10 astronauts did everything except land on the Moon, but that was over a relatively short time and distance. For a long mission to Mars, would such a trial run be considered? A spacecraft could be used for interplanetary travel *and* a planetary landing; Mars Pathfinder, for example, carried the Sojourner rover.

But there are other aspects of the mission to consider, such as the prevention of contamination of Mars by microbes from Earth, and the return journey home. The former might be satisfied by having astronauts sited on Phobos to control robots on Mars. Teleoperation would involve only a very small time-delay (1 second or less) and even allow virtual-reality control. Phobos might also prove to be very beneficial as a source of water for the manufacture of propellant for the return journey, a weather station for dust-storm warnings, or a test-site for asteroid-mining technology. Moreover, due to its very low gravity and lack of atmosphere, it is comparatively easy to land on it and launch from it.

## COMMUNICATIONS

The disadvantage of working in Mars orbit, or even on Phobos, is the risk of exposure to radiation – and the astronauts might be in orbit for fifteen months before the planets are favourably placed for the return journey. If it is decided to land on the surface, communications must be taken into consideration; and if it has to be confirmed by mission control on Earth, the crew in orbit around Mars might be waiting up to 40 minutes for the reply. During that time, any number of situations might force the hand of the crew, and whether it is a scripted or an emergency landing, or even an abort, it will be essential to have a communications network in place to provide as much information as possible.

### TDRS, or ground stations
In the Apollo days, NASA relied for communications on the Manned Space Flight Network (MSFN) of tracking stations, aircraft and ships, plus the Deep Space Network (DSN) of three large-dish aerial tracking stations. The Space Shuttle was limited to Earth orbit operations, but it inaugurated a breakthrough in human spaceflight communications: the geostationary communications satellites network Tracking and Data Relay Satellite System (TDRSS). Data was received from the TDRSS by a ground station at White Sands, New Mexico, where dish antennae pointed at satellites low on the horizon in the east and west. Subsequently, the Advanced TDRS network was created using a new set of satellites launched on

expendable rockets. TDRS even makes possible communication with a spacecraft at the time of black-out during re-entry into Earth's atmosphere, by using viewing angles to the satellites to communicate through the wake of a plasma shroud from above and behind the spacecraft. TDRSS users include the Space Shuttle, the ISS, Earth-orbiting science satellites, and Department of Defense satellites.

The Deep Space Network has already been used for robotic planetary missions, and the reinitiation of human missions beyond Earth orbit may reopen plans for an orbital version. The DSN is operated out of the Jet Propulsion Laboratory in Pasadena, California, and in turn by the California Institute of Technology, for NASA. Its three antenna complexes are sited approximately 120° apart on three continents – at the US Army's Fort Irwin reservation, 72 km from Barstow in the Mojave Desert, California (Goldstone); near the Tidbinbilla Nature Reserve in Australia, 40 kilometres from the capital city (Canberra); and at Robledo de Chavela in Spain, 60 km from the capital city (Madrid). These provide constant communication with distant planetary spacecraft as the Earth turns. All three sites are semi-mountainous, in bowls which act as interference shields. The complex is also used by other countries – notably Europe and Japan – and is supplemented by Japanese and European tracking stations. At the end of 2003 and in 2004, time allocations for the DSN, for communicating with particular spaceprobes, were fully utilised.[17]

### Areocentric satellites
The rotation period of Mars is about 24.5 hours – slightly longer than Earth's day. This causes a loss of communication between Earth and bases or vehicles on the martian ground for 12–14 hours each day.[4] Data-relay satellites could be placed in equatorial orbit around Mars, orbiting synchronously with the rotation of the planet. The equivalent at Earth is geostationary Earth orbit (GEO), used by communications satellites such as direct-broadcast TV satellites so that reception dishes outside offices or in gardens can be pointed at a fixed position in the sky. A trio of areocentric Mars orbit satellites, cross-linked by data relay, would allow constant communications between Earth and any craft on or near Mars, but satellites in orbit around Mars could also perform a variety of other duties, including navigation, weather monitoring and solar activity monitoring. Three equidistantly-spaced satellites would allow constant coverage for all of these functions, and multifunctional satellites would be able to cover several of these uses, reducing the need to take or send groups of satellites for each purpose.

### Solar conjunction interference
When the Sun lies between Earth and Mars, radio signals between the two planets have to pass through the Sun. This causes a dramatic signal deterioration and a huge increase in levels of static, which would persist for part of a month each martian year. The communications link performance is determined by the signal-to-noise ratio, which can be controlled by using different frequencies, larger or more powerful transmitters, and larger receivers on Earth.[4] Modern planetary radio systems use Ka-band transmitters (37–38 GHz).

One solution to the problem of sending a high volume of digital data from Mars to Earth is laser communications (lasercom) – a bright laser pulsed to encode the information, which is received via a telescope. However, adverse weather – including heavy cloud on Earth, or clouds and especially dust-storms on Mars – can interfere with the signal. On Earth, light pollution (a major problem which afflicts optical astronomy) could also interfere with signal reception.

The communication delay time between Mars and Earth is 10–20 minutes – 20–40 minutes for a two-way conversation – and there have been studies into the feasibility of having a mission controller on Mars.[13] The controller would be among the crew of the mother-craft in orbit or in the habitation module on the surface, and would enable the crew on Mars to be as independent as possible. However, a system of communications satellites would still be required if EVA operations were underway some distance away or if teleoperated robots were in operation on the other side of the planet.

**Spacecraft communications**

Video of the first landing will not be seen until at least twenty minutes later (the 1-second time-lag during the Apollo missions seems trivial by comparison), and the conspiracy theorists would probably have a field day with the coverage not being 'live'. But it is physics, and cannot be avoided.

The crew of the Apollo Lunar Module soon erected a high-gain antenna on the Moon during an EVA after landing, and the same may be true on Mars. Signals could also be sent via a large communications dish on the orbiting mother-craft, rather than direct to Earth.

However, all will not be lost if, for example, the mother-craft's high-gain antenna malfunctions. The crew could carry out the repairs on EVA, and the mission could still proceed with only a low gain-antenna (as happened during the Galileo mission to Jupiter), although the data transmission rate would be drastically reduced.

All of the communications systems would have to be functioning properly before any landing could be attempted and before any surface EVA. But, assuming all was well and the lander was safely down, by whatever method, the first surface excursion of this initial manned landing would be the equivalent of Neil Armstrong's 'giant leap'. Marswalk One would then begin, though probably not for a few days.

**REFERENCES**

1  Interviews with Bo Maxwell, 5 and 16 December 2003.
2  Interviews with Alex Ellery, 10 October 2003 and 4 February 2004.
3  Benton C. Clark, *Crew Activities, Science and Hazards of Manned Missions to Mars*, Planetary Sciences Laboratory, Martin Marietta Astronautics, Denver, Colorado, 39th Congress IAF, Bangalore, India, 8–15 October 1988, IAF-88-403.
4  *Report of the 90-day Study on Human Exploration of the Moon and Mars*, NASA.
5  *Radiation Hazards to Crews of Interplanetary Flights*, Space Studies Board, National Research Council.

6 Gilles Clément, *Fundamentals of Space Medicine*.
7 *Manned Lunar, Asteroid and Mars Missions*, SAIC for The Planetary Society.
8 *Exploring the Moon and Mars: Choices for the Nation*, OTA-ISC-502, 1991.
9 *Radiobiological Factors in Manned Spaceflight*, SSB NRC, 1967.
10 *Radiation Protection Guides and Constraints for Space-Mission and Vehicle-Design Studies Involving Nuclear Systems*, SSB NRC.
11 *Guidance on Radiation Received in Space Activities*, National Council on Radiation Protection and Measurements, 1989.
12 Dr Michael Reichart and Dr Wolfgang Seboldt, 'Vision Mars: the First Manned Flight to our Red Neighbour', in *Towards Mars!*
13 Interview with Michael Duke, 28 January 2004.
14 Geoffrey A. Landis, 'Humans and Robots: Synergy in Planetary Exploration', *Acta Astronautica*.
15 David J. Shayler, *Skylab*; David M. Harland, *The Space Shuttle*; David M. Harland, *The Mir Space Station*.
16 Carol R. Stoker, *Science Strategy for the Human Exploration of Mars*, AAS 95-493.
17 *The Deep Space Network*, JPL 400-333.

# Evolution of a Marswalk

For decades prior to President Kennedy's commitment to the Moon, there were countless suggestions about what the first men would do once they arrived there. Once Apollo became an authorised programme tasked with landing men on the Moon by 1970, the eight years between Kennedy's commitment in May 1961 and the Apollo 11 landing in July 1969 were full of proposals, plans, studies, ideas and suggestions for what lunar landing missions could or should attempt. Almost two generations later, history looks set to repeat itself. For decades we have pondered manned flights to Mars, including surface exploration, but we have yet to define exactly what the first landing will achieve, simply because we have no clear idea of when we will go, how to will get there, or how long we will be staying. So, defining exactly what the first Mars explorers will do is a leap into the unknown – prone to changes of technology and direction by the time we first walk on another planet. However, we can trace the ideas and suggestions from the history of early Mars exploration plans, and present a scenario that could probably be attempted.

## CONCEPTS FOR A MARSWALK

By the 1950s, serious thought was being given to exploring space, the Moon, and Mars. This generated a wealth of written proposals on exactly how people would journey to the Red Planet and what they would do when they arrive.

### Das Marsprojekt

In 1952, some of the ideas and plans evolved by Wernher von Braun for future exploration of Mars by human crews were published under the title *Das Marsprojekt* (*The Mars Project*).[1] This outlined a vast undertaking, utilising ten 4,000-ton spacecraft and a crew complement of up to seventy people. Landing would be achieved by a trio of winged landing craft, with the first using skids to land at the site (chosen before Earth departure) in either of the polar regions, where the smoothest landing places were then thought to be. The crew would then leave their lander at the pole and conduct a 4,000-mile expedition to the chosen base camp at an equatorial

Depending upon the final profile, initial activities on the surface will include making the landing area safe and establishing a base. This concept has been a focal point for the ever-changing evolution of a Marswalk. (Courtesy Pat Rawlings.)

site previously identified from orbit. Once the base is established, the other two landers (with wheels) would descend from orbit with the majority of the crew, leaving a caretaker crew in orbit, and the team would then create an inflatable habitat to support up to 400 days of surface operations. The surface expedition would include professional scientists to increase the scientific yield, and a fleet of tracked surface-vehicles to extend the operational range of the team. It would be a daring adventure for the first men on Mars.

Two years later, von Braun's ideas were amended and included in his famous article 'Can we go to Mars?', published in *Colliers* in 1954.[2] Then, in 1956 he published, in cooperation with Willy Ley, a revised version of his thoughts on initial surface exploration – this time proposing a team of nine crew-members descending straight to an equatorial site. Immediately after landing, they would prepare the vehicle for emergency lift-off (as did the Apollo landing crews) by almost reassembling their spacecraft, and would then inflate habitable hemispherical living quarters – all after spending eight months in microgravity. The nine men would spend a year on the surface before returning home to a heroes' welcome.[3]

**Cosmonauts on Mars by 1969?**
In the early 1960s, under the leadership of OKB-1 Designer Mikhail K. Tikhonravov, a group of talented Soviet engineers, headed by Konstantin

Feoktistov, conducted a study into sending the first cosmonauts to Mars. The Soviets had had remarkable success in their early programme, with notable firsts, and this inspired the designers to reason that the first person on Mars should be Soviet. Tikhonravov stated that in his department there would be no place for an engineer who did not strive to understand that a cosmonaut should stand on Mars by 1969 – and many of the engineers hoped it could be achieved by 1966! In Tikhonravov's plan, a six-man expedition would arrive at Mars and study the planet from orbit. Two cosmonauts would then use a 'research manned spacecraft' to explore Phobos and Deimos, including the retrieval of geological samples. Later, a crew of three would land on the surface of Mars and use a 'vehicle' of four linked, wheeled 'carriages' to travel from pole to pole.[4]

**Early NASA plans, 1960–64**

NASA was formed from the former National Advisory Committee of Aeronautics in October 1958, and while its Man-in-Space programme soon focused on Project Mercury to place an American in Earth orbit for a day, and future circumlunar missions under Apollo, some early studies on future manned flights to Mars were also conducted. In the 1960 study by the NASA Lewis Research Center, a 420-day journey to Mars (to begin in May 1971) included a 40-day stay on the surface[5] by two of the seven astronauts proposed for the mission. In 1962, a study by Ernst Stuhlinger's group, within the Army Ballistic Missile Agency at Huntsville, proposed a combination of five Mars-bound spacecraft, each carrying a crew of three (fifteen astronauts). There would be three landing vehicles, capable of being either an unmanned cargo/crew rescue vehicle or a manned transport craft, which would support a small surface exploration crew taken from the total crew complement. Under normal situations, the unmanned lander would be followed by the manned craft in a surface mission lasting 29 days.[6]

The following year, at the Manned Planetary Mission Technology Conference held at NASA's Lewis Research Center between 21 and 23 May 1963, representatives of the MSC Spacecraft Technology Division presented the results of Mars exploration studies conducted at Houston. In one of the designs, using Apollo/Saturn-derived hardware, a crew would land on the surface in a Mars Excursion Module (MEM), having left the 'direct' support craft used for the flight from Earth. They would spend the next 40 days exploring the surface while awaiting the arrival of the fly-by craft, which would be unmanned and used for the journey home following ascent from the surface and a two-day rendezvous and docking phase. This mission would be extremely risky, as the crew would need to rely on the rendezvous with the return spacecraft as it flew by Mars. In another mission design, a separate lander would be dispatched from an orbiting mother-craft (similar to the Apollo design), and remain on the surface for 10–40 days, before lift-off and rendezvous with the orbiting mother-craft for the journey home.

These studies were further evaluated by the Philco (Ford) Aeronautical Company in a presentation at the third American Institute for Astronautics and Aeronautics (AIAA) Manned Space Flight Conference, held in Houston in November 1964. In this plan, a three-man crew would land a small MEM on Mars, and stay there for

10–40 days. The crew would include the Captain/Scientific Aide, a First Officer/ Geologist, and a Second Officer/Biologist. Access between the lander and the surface would be enabled by means of a cylindrical airlock that could be lowered from the MEM base to the surface (similar to an elevator car). The EVA programme was envisaged as 16 man-hours per day – which probably entailed two men performing 8 hours of surface activities and the third remaining inside, on a rotational system, including a day off for rest and planning. A communications antenna would be erected for direct transmissions to Earth, and surface operations would include the collection of 'biological specimens'.[7]

### The golden era, 1968–71

By 1968, more refined studies were being proposed in light of new (although rather limited) data from Mariner 4 and with the impending Mariner 6/7 missions. In January 1968, the Boeing Company published the results of a NASA-contracted fourteen-month study,[8] which was by far the most detailed study of manned exploration of Mars up to that time. The lander – designed by North American Rockwell[9] – would descend from the mother-craft, which was designed by Boeing. The design of the lander varied from a vehicle to support two astronauts for four days on the surface to one that could support four astronauts for up to 30 days, and featured a ladder for crew descent and ascent and a ramp for logistics and surface-vehicle deployment.

On 4 August 1969, Wernher von Braun presented the Space Task Group with his ideas for manned exploration of Mars in the early 1980s. He envisaged a pair of

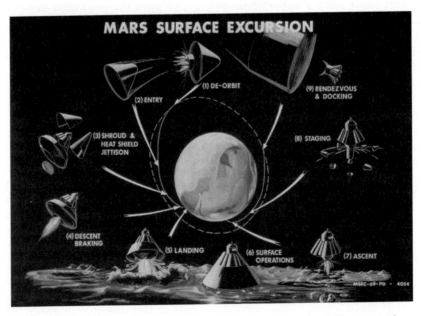

The sequence of Mars operations from Wernher von Braun's 1969 presentation.

spacecraft, each carrying a complement of six astronauts and one three-man lander. At Mars, one lander would make the descent, while the other would act as a possible rescue vehicle. Upon landing, the crew would spend 30–60 days on the surface, with the following primary objectives:

- The collection of geological samples for a better understanding of the planet's origin and development.
- A search for signs of life, and observation of the 'behaviour of terrestrial life-forms transplanted to the martian environment.'
- Drilling for subterranean water that could be used to produce return-journey fuel for later missions.
- The return of equipment and up to 900 lbs of geological samples.

Details of proposed surface activities included conducting geophysical observations and gathering data on the gravitational and magnetic fields, as well as extracting core samples and conducting heat-flow and seismic experiments to determine the internal composition of the planet.[10]

The shape of the 1969 design for the Mars Excursion Module (MEM) resembled the cone of the Apollo Command Module. Von Braun noted that the Mars orbital spacecraft resembled the Apollo 11 Command Module *Columbia*, and the MEM was like the Apollo Lunar Module *Eagle*. The 'Apollo' shape of the vehicle was designed for protection during descent through the atmosphere and support during the surface stay, and would also act as a launch pad for the ascent back to orbit. Designed to support the three-man landing crew, it would act as the crew return

Astronauts conducting the first Mars EVA, as portrayed in von Braun's presentation.

vehicle and ferry for returning scientific data and samples to the orbiting mother-craft.

In addition to a laboratory and living quarters for a 30–60-day stay on the surface, the MEM (like its precursor, the Apollo LM) incorporated ascent and descent stages. The ascent stage was also the main control area supporting the crew during undocked orbital flight, entry, descent and landing, and during ascent, rendezvous and docking, with provision for the ascent engine and for propellant and consumables storage tanks. The living quarters and the laboratory for use on the surface were in the descent stage, together with the descent engine propellant tanks, landing gear and outer heat shield for aerodynamic entry, and support systems and consumables for the duration of the stay on the surface. In this proposal, the crew would have a one-man Mars rover (stored in a 'hangar' facility) for mobility during the EVAs. All descent equipment was to be left on the surface (as with Apollo), and the design was such that a single astronaut would have been capable of landing on Mars or bringing a stranded crew back to orbit. At departure from the orbital craft, the fully loaded MEM had a mass of 95,000 lbs and a maximum base diameter of about 30 feet.[11]

Von Braun suggested that the first landing on Mars would in many ways resemble the activities on Apollo 11, but with a much longer stay on the surface (30–60 days rather than 36 hours), allowing for a far more extensive programme of observations, experiments and surface exploration to attain a broader range of mission goals. Use of the Mars Roving Vehicle (MRV) would allow exploration beyond the immediate range of the lander, and the availability of a scientific laboratory would allow experiments to be conducted in the MEM as well as on the surface. During the period of surface activities, the three astronauts in the orbital spacecraft and the six in the second spacecraft would continue their own experimental programmes,

A cutaway of the Mars Excursion Module (MEM) revealing internal arrangements.

observations and monitoring of surface operations, as well as maintaining the system performance of their own craft.

A second landing of the same duration would follow the return of the first team, and the departure from orbit would take place 80–90 days after arrival via a Venus fly-by. This was a plan for a single, dual-landing mission, departing from Earth in 1981 and achieving the landing on Mars during 1982. It utilised the experiences gained and derived the hardware from Apollo, as well as nuclear technology and new mission-specific hardware. It was expected to be followed by the creation of a twelve-man temporary Mars research base around 1985, and a semi-permanent base at the planet, with forty-eight men on the surface and twenty-four men in orbit, by 1988/89 – 20 years after Armstrong stepped onto the Moon. It was a bold and far-reaching plan – part of a large programme that included the development of space stations, bases on the Moon, and planetary exploration.

These studies came at a time when NASA was recovering from the Apollo 1 pad fire and preparing to mount the final assault on the Moon, leading to the historic landing of Apollo 11 in July 1969. The first lunar landing surface exploration phase had been amended to a short 2-hour excursion with no extended geological traverses, plus deployment of limited surface experiments. The idea was to focus on the safety of the astronauts and on achieving the primary goal of landing on the Moon and returning to Earth. More detailed science would be conducted on later missions. But at NASA, plans for what would follow Apollo were already suffering from budget restrictions caused by several managerial, social, domestic, political and international factors.

The 'grand plan' for post-Apollo operations in Earth orbit, at the Moon and out to Mars – which was marketed as the next logical step in space in 1969 – was essentially dead by 1970, with only the Shuttle surviving the budgetary and political axe. However, in October 1969 the Planetary Missions Requirements Group (PMRG) was established to define manned Mars mission concepts in light of the plan proposed by the Space Task Group (STG). Although budget limitations dictated the activities of this group, low level studies continued, and the 1971 study[9] included components launched by the Shuttle to fabricate a chemical or nuclear propulsion vehicle with a crew of five. In this study, a team of three would land in their MEM, leaving two in orbit to monitor operations. The study also took into account the experience from the Apollo EVAs and the results obtained by Mariner 6 and Mariner 7, to evolve a more detailed EVA plan.

The stay on the surface would last for 45 days, and operations would be carried out by two astronauts performing surface activities with the third remaining inside the MEM (probably acting as EVA choreographer and CapCom). Wearing hard-upper-shelled aluminium pressure suits, the astronauts conducting surface exploration would use two unpressurised and electrically powered rovers derived from the LRV developed for the last Apollo missions. They would be capable of a top speed of 10 mph, and by using two vehicles the astronauts would increase their potential exploration area while still observing the safety rules of the mission. Following experience gained during the development of the Apollo LRV, the Mars mission would have a redundancy capability, with each man driving a rover so that in the

Mars lift-off. The MEM ascent stage begins the first leg of the long flight home.

event of a breakdown they could both ride the second rover back to the lander and tow the afflicted vehicle, after which they could either repair it or cannibalise it for spares. This alleviated the Apollo 'walk-back' rule, by which the distance from the Apollo LM was dependent on the capacity of the pressure garment consumables (water and oxygen) rather than the ability of the crew-member.

During the 45-day surface stay, two 36-hour traverses would be undertaken on day 15/16 and day 30/31, with an excursion of up to 152 miles. The two astronauts would spend the 'night' sleeping in their pressure suits on the rovers. The report was not forthcoming on how much sleep was expected during this 'night under the stars', but it would not be much. Experience from Apollo showed that the early crews did not have much sleep inside the cramped and basic LM, and so any attempt to catch a nap while wearing a pressure suit and sitting on an open rover would be, at best, uncomfortable. Presumably, life support, eating and drinking, and waste collection, would be possible inside the suit system. In addition to gathering geological samples the astronauts would actively search for signs of 'even elementary life'.

Less than a year after this report was published, Mariner 9 was returning stunning photographs of craters and dust-storms; and once again the image of Mars, together with ideas for human missions to the planet, were changed.


## MARS IN A NEW LIGHT

In 1976 the surface views and staggering orbital photography returned by the highly successful Viking probes yet again revealed the planet in a new light. Information on atmospheric composition gave rise to a drive to design a Mars exploration mission

An artist's impression of orbital rendezvous of the ascent stage and the main spacecraft, from the 1969 proposal.

that would essentially 'live off the land' by extracting gases from the atmosphere and converting it into fuel for the return journey or to sustain surface expeditions for longer periods. But talk of manned Mars exploration missions was placed in the background as NASA concentrated on orbiting the Space Shuttle. Two weeks after the first flight of the Shuttle in April 1981, the first Case for Mars conference suggested that no astronaut would land on Mars, but that a crew of six to eight would control a sample-return vehicle from orbit, and that two of the astronauts would land on Phobos, controlling a fleet of ten to twenty automated Mars Rovers during a two- to six-month stay at Mars. This scenario was published as part of the Office of Exploration's 1988 'Beyond Earth's Boundaries' report to the NASA Administrator. The report stated that the two astronauts would spend twenty days on Phobos, and also indicated that a subsequent expedition to Mars would include a 20-day stay on the surface.

In 1983 the Planetary Society commissioned its own piloted-mission study, in which three of four crew-members would land on the surface for a one-month stay and return to orbit with 400 kg of rock samples. In 1984 the Case for Mars II conference suggested a rotating exploratory crew, with a permanent base supporting a surface crew for two years and rotated with a fresh team every twenty-four months.

In the 1987 Ride Report for the future of American leadership in space after the year 2000[12] it was suggested that robotic missions would precede manned flights to Mars, as would research into long-duration manned spaceflights onboard the ISS. The first stays on the surface would be of 10–20 days during a 30–45-day orbital stay in a mission lasting one year. These preliminary missions would have the primary aim of exploring and identifying potential sites to set up a research outpost and gain experience of working on Mars, as well as building up a database on the risks and experiences of long-duration deep-space missions. After ten years, a manned research base would be established.

The 90-day Space Exploration Initiative (SEI) study, completed in 1989, forecast a four-astronaut team to journey to Mars for a 30-day stay on the surface in 2016. Various options were presented, balancing the cost and progress to achieve a 30-day stay, followed by a 60-day stay two years later, and then a 90-day stay, leading to a permanent outpost supporting the first 600-day surface tour of duty.

At the same time as the 90-day study, an evaluation was presented by engineers at the Lawrence Livermore National Laboratory, as part of a defence and advanced technology programme which expanded the work at the national laboratories at a time when the demise of the Cold War threatened their budgets. This plan proposed the use of Apollo-type hardware to set up an Earth-orbital space station, a refuelling station, a lunar base, and a manned flight to Mars, where the first crew would explore Phobos and Deimos and then spend 399 days in an inflatable habitat on the surface of Mars, operating a mining system to extract water for use in the manufacture of propellants for feeding rocket-powered surface-hoppers.

With these proposals, a growing desire to cooperate with the Soviets in manned and unmanned spaceflight, and a growing need to redesign the over-complicated and over-budget US Freedom space station, an SEI Synthesis group, headed by former astronaut Tom Stafford, presented another high-profile document on the prospect of the future direction of US spaceflight. The report – entitled 'America at the Threshold', and published in May 1991 – emphasised a Mars transportation infrastructure (without a space station), a return to the Moon that included Mars simulations, and the first Mars landing crew spending 30 days on the surface, testing systems and exploring the landing area. On a second landing four years later, the crew would spend 600 days on the surface.

Despite difficulties in maintaining their space programme in the 1980s, the Soviets still dreamed of flying cosmonauts to Mars. In 1987, NPO Energiya published a report on a study into sending cosmonauts to Mars, during which two cosmonauts would spend a week there.[13] Seven years later, *Zemlya I Vselennaya* (*Earth and Universe*) published, in its November–December 1994 issue, an extract from a forthcoming book *Kosmicheskiye Orbity* (*Space Orbits*) by S.P. Umanskiy, regarding the prospect of manned Russian spaceflights to Mars. In his explanation of the use of electric rocket engines (ERE) and solar cells developed at NPO Energiya, Umanskiy proposed that two members of a four-man crew would stay on Mars for seven days.[14]

## MARS DIRECT

Studies into the Mars Direct approach have been conducted since 1992, and over the past decade have generated much interest, including the simulations conducted by the Mars Society on Devon Island and other locations. These were major surface operations involving pin-point landings and long stays of up to 500 days, with

| Report year | Crew numbers | Surface time (days) | EVAs and activities |
|---|---|---|---|
| 1952 | 50? | > 400 | A 4,000-mile trek from the pole to the equator, followed by local EVAs around a central base. |
| 1956 | 9 | 365 | |
| 1960 | 2 | 40 | |
| 1960 | 3 | ? | A Russian plan for two cosmonauts to survey the moons of Mars, then a three-man team to traverse Mars from pole to pole using a pressurised and unpressurised train of vehicles. |
| 1962 | 3? | 29 | |
| 1963 | 3? | 10–40 | |
| 1964 | 3 | 10–40 | Two-man EVAs totalling 16 man-hours per day. |
| 1968 | 2 | 4 | |
| 1968 | 4 | 30 | |
| 1969 | 3 | 30–60 | Search for life and return 900 lbs of samples and equipment. |
| 1971 | 3 | 45 | Two-man EVAs, including two 36-hour 152-mile 'overnight' traverses in two rovers; exploration area of up to 8,000 square miles. |
| 1983 | 3–4 | 30 | Return with 400 kg of geological samples, |
| 1987 | ? | 10–20 | |
| 1987 | 2 | 7 | A Russian study, |
| 1988 | ? | 20 | (Preceded by a precursor mission that included a 20-day stay on Phobos but not a descent to Mars). |
| 1989 | ? | 30 | Follow-up mission of 60, 90 and 600 days. |
| 1989 | ? | 399 | Mining water from the surface for use in propellant manufacturing, and geological sampling. |
| 1989 | 2 × 4 | 2 × 45 | Two separate landings during one mission supporting 300 man-hours of EVA for each team of four astronauts. |
| 1991 | ? | 90 | Exploring and testing new technologies and systems to support a second mission of 600 days on the surface. |
| 1992 | ? | 500 | Set up a surface base after the first landing; 600-km round-trip traverse. |
| 1993 | ? | 600 | 500-mile traverse over 10-day periods. |
| 1993/4–99 | 6 | 500 | Each astronaut to perform a maximum of two EVAs every three days – a total of 2,000 individual EVA periods (333 EVAs for each astronaut). |

extended-duration traverses of up to 600 miles on a round-trip. Depending on the success of this expedition, a two-year follow-on mission pattern would be established to create a number of small bases across the surface.

In support of these studies, in August 1992 a NASA Mars Exploration Study Team was formed. In May 1993 it produced a report entitled 'The Design Reference Mission' (DRM), which proposed that an unmanned habitat and cargo lander would land before the first human crew in the next launch opportunity. The crew would stay for 600 days and use pressurised rovers to carry out a ten-day 500-km traverse from the Mars-1 base.

In the late 1990s – amid the successes and failures of the unmanned probes to Mars and the commencement of ISS operations – discussions continued as to what should be done in the approach to Mars – not only for the first crew, but also over the longer term. Technical difficulties, together with costs, and the distance of Mars, contributed to indications of a manned Mars mission 'some time in the future' and a 'will they?/won't they?' guessing game. In almost forty years, the general consensus had changed concerning the duration of the first landing crew's stay on Mars. The table on the previous page is a selected study timeline of surface operation proposals over five decades.

It is interesting to view the timelines of the surface durations in these proposals. In the early years they are more than a year, based on very limited information about the planet, whereas for most of the period 1960–90 a period of around one month was standard. More recent reports from the 1990s increased this to around two years, with most of the consumables carried to or produced on the surface. The key of course, is at what point in the Earth–Mars orbital relationship the mission is undertaken, and when the expedition crew would descend to begin exploration of the surface. It is reasonable to assume that, for safety, a short seven-day period of surface exploration would be based upon precursor missions and past experience of lunar exploration (Apollo or its successor), and a 30-day period of exploration would be the most practical to include both a scientific return and a safety overhead until more could be learned from operations conducted on the surface. Without the availability of redundancy, consumables and crew rescue options, a 300- to 600-day stay on the planet for the first landing would be a huge risk, especially if an Earth–Mars infrastructure was not in place.

**How long to stay?**
Once the crew has successfully achieved the first manned landing on Mars and established that it is safe to stay (at least for a while), it must next be decided how long they will stay there. During the Apollo missions, several 'stay or no stay' decisions were made during the first minutes and hours after landing, as the spacecraft's systems and status were checked by both the astronauts and the team of flight controllers on Earth. But there was only a 2-second gap in communications, the lone CSM pilot was still flying overhead, and Earth was just three days away, so there was an element of added safety for the astronauts on the Moon. At Mars, conditions will be completely different. Communications would initially be with the orbital element of the main spacecraft, but as that passed below the horizon, communication would have to take place via an orbital communications satellite. Direct communication with Earth would involve a delay of more than 16 minutes at very best, and the landing crew would be isolated for some of the time. The mother-craft would be in either low orbit around Mars, like Mars Odyssey or in a highly elliptical orbit, like Mars Express. The orbit might be near-equatorial or near-polar, or it might even be an areocentric orbit (equivalent to a geocentric orbit around Earth). Switching between these different orbits would involve significant usage of

propellant; but it might be possible, depending on the propulsion method used in delivering the mission to Mars. Upon arrival, the release of a few orbital communications satellites would produce a network to provide continuous radio contact between the mother-craft and the lander. Three communications satellites in areocentric orbit would probably be sufficient.

The duration of the stay on the surface would also depend on the method chosen to reach the planet and exactly when in the orbital cycle the landing is accomplished. A few days or weeks on the surface could easily stretch to several months, depending on planetary alignment. Regardless of the duration on the surface, however, there are several tasks and operations that would form the basic objectives of the surface crew.

On arrival at Mars, the expedition would probably have despatched unmanned robotic spacecraft into orbit for communications and additional photographic cover and dust-storm and solar activity alerts. Others would be sent to the surface at a wide variety of locations controlled from the orbiter or from the lander, as precursor survey missions prior to the first crew exit while the crew acclimatised between the long period of weightlessness or partial gravity during the interplanetary flight and the $\frac{1}{3}$ gravity on the surface. It is unlikely that a surface exit by the crew would be undertaken within the first few hours – or possibly even days – after landing, despite the temptation due to being so close, and the spectacular views outside the windows. Unless mission status required an early (though not emergency) lift-off, the crew would rest, and adapt to their new environment for a few days before venturing outside. During Apollo there was a contingency for short, one-person EVAs close to the lander if restrictions were enforced on nominal two-person EVAs,[15] and it seems reasonable to assume that such an option would be available on the early Mars landings to at least secure a small sample by 'human hand', after having taken so much time and effort. If no life-threatening emergency situation presented itself, but a less critical event or situation arose preventing a prolonged stay, then the crews might opt for an abbreviated back-up contingency EVA scenario supported by robotic technology and facilities, before blasting off and heading for orbit.

When the first astronauts step onto the surface for the first EVA, there will inevitably be personal comments and observation. This is human nature, but it is also useful for scientific interpretation. The Apollo astronauts demonstrated that, with proper training, even pilots can become competent geologists; but this was a reflection of the early programme, and following the return to the Moon and the use of science officers on the ISS it is reasonable to assume that scientific specialists would be part of the crew and would be cross-trained in other disciplines for support and back-up. In the first EVA or EVAs on the surface, it is quite probable that after gathering contingency samples, evaluating surface mobility to determine the best method of moving across the surface, and testing the mobility of the Mars EVA suit, short traverses close to the lander would include the deployment of small surface experiments (ALSEP style – and possibly termed Mars Surface Experiment Package (MSEP)).

For the very first EVA on Mars, a simpler MSEP – analogous to the EASEP (the Early ALSEP on Apollo 11) – would be used. This is not a consequence of delays in the preparation of equipment (which led to EASEP on Apollo 11), but an attempt to keep the first Marswalk as basic as possible. A full MSEP would be carried on later EVAs on the same mission, but for the first Marswalk the scientific instruments might be limited to the deployment of a weather station and a passive seismometer to supplement the field geology undertaken in that location. There would be a range of photographic and TV broadcasts and a number of commemorative broadcasts and ceremonies to mark such an historic achievement, but with these out of the way, the crew could then settle down to the more extensive EVA programme for the duration of their stay on the surface.

The duration of the first EVA, by two or more astronauts, would be about 2–8 hours, depending on the consumables of the suit and the ability of the astronauts to follow the programme. For the first EVA, this could probably be quite hurried – packed with activities to fill the timeline, but dependent upon whether the planned longer stay was authorised.

**Hazards for EVA on Mars**
Having completed the journey from Earth to Mars and survived the landing, the difficulties of exploring the surface do not diminish by being on the planet. Some of the hazards awaiting EVA crews on Mars were highlighted in the 1988 IAF paper by Benton Clark.[16] The terrain is completely different from the Moon, simply because of Mars' atmosphere, and the weathering cycles produce hazards such as planet-wide winds blowing dust and sharp rocks across the planet, which could easily tear the pressure garment and result in a quick but painful death.

Scenes of astronauts falling on the Moon are often seen in a humorous vein, but with the danger of a damaged suit always present. The $\frac{1}{6}$ gravity allowed most of the astronauts to react quickly to cushion their fall and bring themselves upright fairly

Iceland has many analogies to Mars. It is a land of 'Fire and Ice' with both active volcanoes and glaciation. Mars also appears to have the remains of both volcanoes and glaciation on its surface today. (Photograph by Andrew Salmon.)

easily. But on Mars – with gravity $\frac{1}{3}$ that of Earth and 2.3 times that of the Moon – falling will be more hazardous, and the suit, with increased mass, will probably be more cumbersome.

## THE 1989 STUDY REPORT

In a 1989 study report prepared for NASA,[17] the study team balanced their own findings with those of the NASA Office of Exploration and JSC planning guidelines, which had planned a 30-day stay in orbit and 20 days on Mars, with 60 man-hours of EVA in five scheduled surface excursions, each of about 6 hours. The team suggested an eighteen-month stay in orbit, allowing for a surface stay of 90 days, and an increase in surface exploration time to 600 man-hours on forty-eight EVAs lasting 2–10 hours each.

The study team deduced that a longer stay on the surface would ease the adaptation from microgravity to $\frac{1}{3}$ g – especially as the crew would probably be unable to function at full peak for about two weeks – although they acknowledged recent reports from the Soviets, who estimated that a faster readaptation might be possible if effective countermeasures were undertaken during the long flight to Mars. In any event, it would probably be impractical to schedule a full EVA on the day of landing, and unwise to embark on arduous 6–8-hour EVAs for the initial few days until the crew acclimatised and were confirmed for a safe stay on the surface.

It was also emphasised that a short stay of around 20 days would create an EVA timeline similar to that of Apollo, during which every minute on the surface counted, and the crews were hustled along to the next objective, station stop or task. Clearly, early EVAs would be prone to losing valuable time as the crew adapted to surface operations and developed an effective working protocol with their equipment and procedures. Reflecting on the experiences of Skylab, the study team realised that to be fully effective, a paced and organised plan would be required to ensure that the maximum return was achieved as easily as possible, with the crew given sufficient time to organise themselves after the prolonged microgravity period, the excitement and challenge of landing, and acclimatisation to $\frac{1}{3}$ g. They would also need time to incorporate the latest updates to the flight plan in consideration of real-time situations. A suggested model for the first EVA outside the lander was also outlined.

The study team proposed, for the first six days, a sequence of readaptation, environmental tests, the deployment of monitoring and other equipment near to the lander/habitat, physical examinations, and planning sessions. During the second week, the crew would complete short and local proximity EVAs as well as the installation and deployment of local equipment such as a more effective communication antennae and a science station, and establish both an agricultural test station and deep drilling operations. The physical examinations would continue on a weekly basis, as would planning sessions, with further expansion from week three. In this plan, the first EVA would not take place until the eighth day on the surface – very different from the Apollo EVAs, which were carried out within hours of landing.

## NASA'S MARS REFERENCE MISSION

As the ideas for human missions to Mars evolved in the early 1990s, the predicted costs escalated. As a result, NASA devised a Reference Mission for the human exploration of Mars that could substantially reduce the expense of the scenarios then under discussion, and would clearly provide a baseline for the fundamental issues. These issues were the scientific objectives of the mission and the capabilities and activities required to achieve these goals. To understand these issues, the chosen approach was to envisage a 'surface mission'. To address this, during 1993 NASA organised a series of workshops and work programmes with an integrated team of engineers and scientists under the leadership of the Exploration Program Office located at JSC in Houston, Texas. The chief result from this series of studies differed from earlier reports by emphasising a longer stay, long-range exploration capabilities from the 'outpost', and a robust system of surface infrastructure.[18]

This Reference Mission drew up an initial simple set of surface capabilities, including the ability to land all the elements for an 'outpost' on one lander, near the earlier unmanned landing system containing the propellant-production hardware for the surface vehicles and for the return to Earth. The crew could then explore the immediate vicinity of the outpost, utilising teleoperated vehicles and pressurised rovers for distances beyond the initial landing area. In the initial report, no details of surface activities, experiments or support tasks were included, but in the ensuing three years the report was updated, modified and expanded to address these shortcomings, finally becoming the published article in 1997. The purpose of the Reference Mission was to provide a mechanism for the human exploration of Mars and an understanding of the nature of these first expeditions, as well as for guidance in the selection of experiments to be carried on future unmanned precursor missions. It proposed a hypothetical scenario with six astronauts spending 500 days on the surface by as early as 2013.

As a result of this expansion of the initial report, in October 1997 a two-day workshop was convened at the Lunar and Planetary Institute to address the surface element of the Reference Mission and the types of scientific and exploratory activity that would be expected.[19] This study prompted the publication of the full Reference Mission document, but the following year (November 1998), the LPI, under the sponsorship of NASA/Human Exploration and Development of Space Enterprise, held a two-day workshop[20] tasked to explore the objectives, desired capabilities and operational requirements for the first human explorers of Mars. This workshop would use the Reference Mission to formulate recommendations to NASA that would contribute to a policy decision. It included world-class specialists in geology and geochemistry, biology and palaeontology, former Apollo astronauts who had explored the Moon and some of the scientists who trained them, Space Shuttle and future ISS astronauts, representatives of the NASA JSC EVA Office, and NASA mission planners. This group would pool their talents and examine in detail the realities of planning and examining a comprehensive programme of sampling and observations to determine the 'compositions, morphology and geological evolution of Mars, and to determine if life ever existed on the planet.'

## WALK THE WALK

During the 1990s, several published papers featured a renewed interest in the human exploration of Mars. Although most of these papers focused on methods of journeying to and returning from the planet, there were some indications of exactly what a crew would do when they arrived there. This was often glossed over, although several studies included the design of pressure suits for such an expedition. In one of these, published in 1999,[21] it was determined that the key to the design of any potential Mars EVA suit was 'the planned length of the mission, the number of EVAs per EMU, and the indigenous resources and physical limitations on Mars.' Using the 1997 NASA Reference Mission guide[22] of 4-hour EVAs with an eventual goal of 8-hour EVAs, this paper assumed a six-person crew on a 500-day maximum surface stay at Mars. Using this baseline, it was assumed that each astronaut would complete an average of two surface EVAs every three days (0.66 EVAs per day per person), which would result in a total of 2,000 individual EVAs during the time on Mars.

Information was presented on the environmental parameters of the Earth, microgravity and Mars; the mass of the suit; system functional requirements, and the effects of the martian environment on various subsystems. The paper also revealed that if each astronaut were to be supplied with only one EVA suit, it would have to withstand at least 333 EVAs totalling between 1,332 hours (333 4-hour EVAs) and 2,664 hours (333 8-hour EVAs) – an exhausting programme for both man and machine. It was suggested that allowance be made for at least two EMUs for each crew-member and an extra day off in the rota, which would result in the following for a 500-day surface stay. The average of 0.66 EVAs per day per person with two EMUs available would result in less wear on each EMU, supporting about 167 EVAs per unit (670–1,330 hours); if a day off was included in the programme, reducing the average to 0.5 EVA per person per day, a single EMU would need to complete 250 EVAs (1,000–2,000 hours), and two EMUs would each complete 125 EVAs (500–1,000 hours).

The duration of EVAs, the number of astronauts, the reusability of hardware, system capability, and the health of the crew, will all contribute to the EVA programme. Unlike Apollo, during which it was known that the crew would each complete up to three 8-hour EVAs on the surface over four days, it is probable that the definitive Mars exploration plan will not be finalised until the crew arrives in orbit around the planet (as in the 1989 NASA Advanced EVA Study), when conditions on the surface, the health of the crew, the status of the vehicles, and updates to the science objectives, could be better determined.

### 2001: a Mars workshop

In January 2001 a NASA-sponsored workshop was organised to open a dialogue between the scientific community – central to planned robotic explorations of Mars – and a group of experts in the systems and technologies critical in any future human exploration missions.[23] This two-day workshop, held at Goddard Spaceflight Center, was sponsored by the NASA Office of Space Flight and the NASA Office of Space Sciences, and was organised and managed by the Lunar and Planetary Institute. It posed questions about the human exploration of Mars compared with purely robotic studies, reviewed what humans could accomplish more effectively than robotic probes, and the potential of a cooperative human/robotic programme. Papers focused on the science opportunities and risks to a human crew, geological investigations, potential science fields required to support human missions, and considerations for EVA at Mars.

R. Fullerton, of the NASA JSC EVA Project Office (XA), outlined the tasks required for humans at Mars to set up and repair the infrastructure (power generation and distribution, and radiation shielding), set up and repair science equipment (surface sensors, drills and rovers), access and study challenging terrain (outcrops, ravines, rock fields and subsurface) during field geological excursions, and rescue crew-members or hardware. He also presented definitive questions (which remain unresolved) concerning the planning of the final programme in which Mars EVAs would be timelined and planned, which would have significant application in the design of hardware and procedures:

- What are the comfortable walking distances and rates for a single 'day' of activity in a pressure garment?
- What are the single day walking distances and rates for a forced (contingency) march in a pressure garment?
- What are the safe return/safe haven facilities, their spacing across the surface, and contents?
- What is the normal duration of an EVA sortie from egress to ingress?
- What is the mandatory duration of consumables margins for nominal and back-up systems?
- What are the normal durations of EVA preparation and post-EVA activities?
- What is the most suitable number of elapsed days after arriving on the surface of Mars before EVA activities commence?
- What are the most likely durations of the first EVAs on Mars?
- During severe dust-storms, what is the minimum distance (or distances) of safe visibility allowed during EVA?
- What are the terrain constraints for a stable footing: slope angles, entering caves, exploring the edges of cliffs or overhangs?
- What are the capabilities of rescue, and by what means (cable, winch, vehicle)?
- Where will injuries be treated: in the field, suited, or unsuited in a safe haven?
- Where do the crew access training materials: in the suit or at a safe haven, or both – and is access full or partial?
- What are the minimum and maximum numbers and locations of EVAs outside for nominal activities and during emergencies?
- What should be the minimal communication and sensor/data definition for voice contact, e-mail access, suit parameters, weather conditions, and navigational positioning?
- Should there be permission granted for recreational or PAO-orientated EVA activities before departure, in transit to Mars, or after arrival?

- What are the guidelines for cable routing and cross-over techniques: are they buried, elevated or positioned under ramps?
- What are the lighting and temperature constraints on the duration, location and distance of each EVA?
- What will be the robotic aid preferences: should they be pressurised or unpressurised, at what range should they operate, and what is the crew and/or cargo capacity of these robotic aids?
- What should be the constraints on recharging suit consumables: should they be avoided for nominal EVA (defining a set duration), or is recharging advisable while on EVA?

**Drawing on four decades of experience**
Although no programme has been defined for demonstrating Mars EVAs with flight hardware and timelines, there has been a long programme of extraterrestrial surface suit tests and simulations that originated with the Moon race in the 1960s.

| | |
|---|---|
| 1960s | Apollo EVA development and US Geological Survey experiences; Soviet manned lunar landing hardware development and simulations; Lunokhod development. |
| 1970s | Apollo operations and USGS experience; Soviet Lunokhod 1 and Lunokhod 2 experience. |
| 1980s | Comparative US suit mobility tests (EMU, Mark III, AX-5) at JSC. |
| 1996 | Comparative US suit mobility tests (A7LB, EMU, Mark III) at JSC. |
| 1997 | US shirt-sleeved geological exercises in Death Valley; lower torso mobility tests with the Mark III suit onboard a KC-135; Haughton-Mars Project, Devon Island, set up by NASA-Ames Research Center. |
| 1998 | Mobility and geological exercises with the Mark III suit at Flagstaff, Arizona; remote site experience in Antarctica. |
| 1999 | US mobility and robotic aid tests (I suit and Russian Marsokhod rover) in the Mojave Desert, California; US mobility tests on the D, I and H suits at JSC; reconnoitre of Devon Island as a future test-site for prototype suit evaluation. |
| Late 1990s | Russian mobility simulations with Orlan-type suits, in Russia, and in cooperation with the Americans in tests of footwear in the US. |
| 2000 | US rover seating tests onboard a KC-135; mobility, geology, drilling, power deploy demonstrations (ATV rover and H/I suits) at JSC and at Flagstaff, Arizona; activation of the first Mars Analogue Research Station – Flashline-MARS, located on Devon Island, and operated by the Mars Society. |
| 2001 | Remote site experience in Antarctica. |
| 2002 | Activation of the second MARS module – the Mars Desert Research Station (MDRS), operated by the Mars Society – in Utah. |
| 2004 | The third MARS facility – Euro-MARS – undergoes tests prior to establishment in the Krafla-Myvatin region of north-east Iceland during 2005. |

2005          Planned set-up of a fourth MARS facility – MARS-Oz – in Arkaroola,
             South Australia.

**2004: a new spirit of discovery**
In January 2004, President George W. Bush initiated his 'vision for US space exploration', committing the nation to a long-term human and robotic programme to explore the Solar System, beginning with a long overdue return to the Moon 'that will ultimately enable future exploration of Mars and other destinations.' The key elements of this latest revision to the nation's space policy stated: 'The timing of human missions to Mars will be based on available budgetary resources, experience and knowledge gained from lunar exploration, discoveries by robotic spacecraft at Mars and other Solar Systems locations, and the development of required technologies and know-how.' This includes the development and demonstration of power generation, propulsion, life support and 'other key capabilities required to support the human exploration of Mars'. No firm timeline was mentioned, and no plans for surface exploration were revealed, but a renewed surge of papers, formal reports, studies, proposals and suggestions will result from this statement over the coming years.

It is still too early to detail just what will happen on the first flight to Mars. This awaits the outline of the first mission. However, despite the constant lack of a firm, clear plan for a journey to Mars, the development of a pressure garment to support initial planetary surface activities has been underway for decades. Even though we do not have a firm commitment to reach for Mars, a system to journey there and return home, or a defined design of spacecraft to support the effort, we can summarise what might be expected from those first EVAs on Mars, based on past experience gained from the Apollo missions, long-duration manned missions in Earth orbit, robotic pioneers, geological explorations, and simulations across our own planet. We can also determine which scientific experiments could be applied to the exploration of Mars, based on our current capabilities and technology. These are the historical and current contributions to the long lead development of the first EVAs on Mars.

**REFERENCES**

1  Wernher von Braun, *The Mars Project*, University of Illinois Press, 1962.
2  Wernher von Braun, 'Can we go to Mars?', *Colliers*, 30 April 1954, 23.
3  Willy Ley and Wernher von Braun, *The Exploration of Mars*, Viking Press, New York, 1956.
4  Leonid Gorshkov and Valeri Lyubinski, 'Our Country's First Martian Project', *Aerospace Courier*, no. 1, January–February 2000, 52–55 (in Russian and English).
5  S. C. Himmel, J. F. Duggan, R. W. Luidens and R. J. Webber, 'A Study of Manned Nuclear-Rocket Missions to Mars', IAS Paper 61-49, presented at the 29th annual meeting of the IAS, January 1961.
6  Ernst Stuhlinger and Joseph King, 'Concept for Manned Mars Exploration with

**Which way to go?**

Pressure garments can be grouped into four categories:

*The soft suit* was chosen for the Gemini, Apollo and Skylab EVA programmes. The astronaut wore a garment consisting of an elasticised bladder and multiple restraint layers made of pliable materials. The problem with this suit was that when inflated it tended to become more rigid, and required some effort to manipulate the arms and legs. These suits also incorporated a separate bladder and a shaping restraint layer for suitable jointing systems. Essentially, wearing it was like living inside a balloon. It was fine for limited use and compact for storage, but was prone to wear and tear from repeated donning, doffing and snagging, and was uncomfortable for extended and repetitive EVAs over a long period, during which, physical exertion was required. The Russian Voskhod and Soyuz EVA suits were also of this type.

*The semi-rigid suit* is a combination of soft limb elements and a metal or plastic upper-body shell. This type of suit can also include joints of laminated fabric,

J.S. Peress climbs into his own deep-sea suit in 1933. (From 'Clothes that clank', with permission.)

although the arms and legs are essentially made of soft fabric. Termed 'hybrid', it has been used on both the Space Shuttle and the Russian space-station (Salyut and Mir) programmes for more than twenty-five years. Its rigidity is an advantage. It maintains a constant inner volume to aid the mobility of the wearer, and the soft limb elements can be folded for ease of storage when not in use. The integrated back-pack of the Russian Orlan suits has also demonstrated its capability for in-flight servicing and repair – a valuable advantage during a long mission away from Earth.

*The rigid or hard suit* is a pressure garment in which at least 70% of the structure is fabricated from rigid material (such as plastic, metal, aluminium or composites). Despite their outward appearance, these suits (resembling modern-day suits of armour) are more mobile than soft or semi-rigid suits due to the constant volume maintained inside. Their main drawback is the difficulty of storing non-collapsible suits inside the habitable compartments of the spacecraft. This type of suit is intended only for EVA operations – unlike the soft (Apollo) and semi-rigid (Shuttle/Orlan) suits that can and have been used for IVA operations.

*The 'skin-tight' suit* employs mechanical counter-pressure (MCP) in a body-hugging elasticised material for protection of the whole body against vacuum, temperature variations, and debris. Also termed 'elastrometeric' or 'space activity' suits, they have only been evaluated in a series of theoretical and preliminary studies.

Additional to these configurations are the studies for the design of helmets, boots and gloves, and integrated life support systems to support EVA operations in space or on the lunar and planetary surfaces. The design of integrated jointing systems, locking devices for gloves, boots and helmets, communication and information systems, and sizing flexibility, all increase the complexity and limitations of the overall suit design and its operating envelope.[2]

### From surface to sea bed

Experience gained during the Apollo programme, and developments in new technologies over the past fifty years, have indicated that suits designed for work on Mars will differ substantially from those used on the Moon. Pure fabric suits are prone to damage and snagging, while hard fabricated suits offer more protection and assist in mobility in alien environments. But there are more issues to address than material selection. At Mars, a 'planetary exploration suit' will be required – but it will not be the first time such equipment has been developed. On Earth, the deep-sea diving 'planetary exploration suit' has been developing over the past three centuries.

Hard suit technology has its origins in the suits of armour of the medieval period. Indeed, during the development of a pressure garment for the Apollo programme, engineers visited the Tower of London and other armouries to examine artefacts of medieval suit technology to determine how their predecessors solved the problem of attaching gloves and helmets to the main components, and how knights could move their arms, elbows, hips and knees when riding, walking and fighting. This 'battle' mobility was also considered for pioneering diving-suit technology from the eighteenth century.

The construction of the first 'semi-atmosphere diving suit' is accredited to John

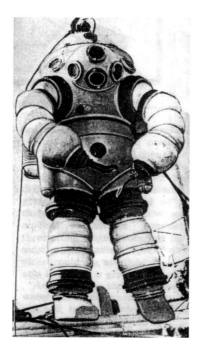

The Galeazzi suit, used by an Italian salvage company in 1931 to recover 7 tons of gold and silver from the *Egypt*, which sank in 1922. (From 'Clothes that clank', with permission.)

Lethbridge, who in 1715 devised a 'diving engine with communication of air', and reportedly used it successfully many times to a depth of 18 metres. The suit included leather cuffs around the bare arms of the diver, but apparently had integrated leggings. Historical accounts of this device are open to debate, but it seems improbable that Lethbridge's claims of spending 34 minutes underwater are exaggerated. In his account of the device, Lethbridge approached the obvious question that has challenged all subsequent explorers of the deep, the stratosphere, space, and eventually Mars: 'How long could I live without air pipes, or communication of air?' He did not use an air-line, but instead employed a pair of bellows, stopped with plugs, immediately before going to the bottom. The device also included a 10-cm diameter, 3-cm thick eye-level viewing port.[3]

More than a century later, in 1838, Englishman W.H. Taylor developed an 'armoured' diving suit featuring ring-stiffened leather bellows used to articulate the knees, shoulders and elbows. But despite being fitted with a 'helmet', the suit lacked gloves and had no protection for the feet, and so it was rather limited for use at depth and duration. It is not known whether the suit was used, or whether it was a design study, but it would certainly have presented problems in retaining pressure at depth, because of the potential failure of the leather bellows under pressure and the short air-hose. The design would have allowed water to enter the suit.[4]

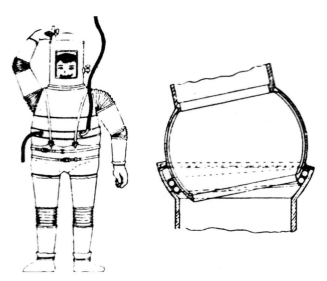

(Left) W.H. Taylor's design, 1838; (right) Neufeldt and Kuhnke's ball-and-socket joint, 1913. (From 'Clothes that clank', with permission.)

In 1856, American Lodner Phillips conceived the first recognised totally enclosed suit, which used ball-and-socket joints to allow movement of the arms and legs. It was jointed at the shoulders, elbows, hips and knees, and the arms had simple manipulator devices which have continued to be a feature of underwater suits. The breathable atmosphere was supplied and evacuated through twin hoses to and from the surface. It is not known if this suit was actually made, but it was the focus for designs that followed over the next fifty years or so.

While hard-hat, weighted boots and fabric suit designs developed, work continued on fully armoured suits and workable joints to assist working at increasing depths. In 1882 the French Carmagnolle brothers patented a suit featuring twenty-two rolling convolute joints that included concentric spheres with a loop of waterproof cloth, designed to keep the joint moveable but watertight. This is generally regarded as the first anthropometric suit. By 1915, American Harry L. Bowdoin had registered a patent for an armoured suit that featured rotary joints; but the joints required oil, and due to the need for constant lubrication the joints could well have dried up and seized. In addition to a viewing port, it featured chest-mounted underwater lamps for improving visibility (current EVA suits feature helmet-mounted lights), and various spacers in the waist, arms and legs so that divers of various heights could adjust and wear the same suit (a feature that is incorporated in current pressure garments). Again, there is no record to show that this suit was made. The design of joints remained a challenge, and in 1913 Neufeldt and Kuhnke developed, in their diving suits, a ball-and-socket design featuring strips of rubber forced against a bearing by water pressure to seal the suit and prevent water leaks. Tests of a second-generation version by the German Navy, to depths of 161 metres, revealed difficulties in moving the jointed limbs, and so it was not completely fail-safe.

During the 1920s, various designs of hard suit were built, developed and tested in Italy, America and Great Britain, and many of them were used for the recovery of artefacts from sunken vessels. In 1931 an Italian salvage company used a variant of the ball-and-socket joint in their Galeazzi suit, which was used in recovering 7 tons of gold from the *Egypt*, which had sunk in 438 feet of water in 1922. Two years later, Englishman J.S. Peress used a variation of the joint in his Tritonia suit during operations on the wreck of the *Lusitania*. This suit featured a layer of liquid trapped between two sections of the joint, which served as lubricant and seal, in the first spherical-type joint. In 1920 another method was developed by McDuffle, in a suit which featured only rotary metal joints – which stretched known foundry fabrication techniques, and were a nightmare.

By the end of the 1930s, several navies had developed a growing interest in the development of ambient pressure diving systems, and the need for deep-sea diving suits diminished. Work progressed slowly on these suits until the 1950s, when American Alfred A. Mikalow developed a ball-and-socket jointed suit that was used to a depth of 99 metres and reportedly had a depth limit of 300 metres. It also featured a selection of interchangeable tools on the end of the arm manipulators, and seven 90-cubic-foot high-pressure cylinders that provided breathing gas and buoyancy and eliminated the need for umbilical connections. It also incorporated hydrophones for communication with the surface, and lights were attached to an arm and on the helmet.

All of these suits were what is currently termed 'atmospheric diving suits', and over the past 100 years they have been developed to support undersea salvage and drilling operations. Since the 1960s they have evolved into advanced diving suits in the oil and gas industries, including the JIM series (UK), the WASP (UK, 1978), SPIDER (UK, 1979), and Newtsuit (Canada, 1985). These have all been very successfully used for exploration and industry in Earth's oceans. Indeed, one of the leading underwater exploration companies – Oceaneering – also has a space division working on the development of EVA equipment and techniques; and a former employee – diver Mike Gernhardt – later became a NASA astronaut and carried out EVAs.

Early JIM suits weighed around 1,100 lbs (including the diver), and in the water had a negative buoyancy of 15–50 lbs. After tests during 1972–75, divers wore these suits in 1976 to dive through holes cut in ice-flows 5 metres thick, and on one dive set a record of 5 hrs 59 min at a maximum depth of 275 metres, in arctic water temperatures recorded at –29° F, although the divers wore thick woollen sweaters to help maintain the internal temperature at 50° F. By 1977, JIM suits had been used on thirty-five dives averaging more than 2 hours at depths of 300–394 metres. The WASP suits featured thrusters, used for mid-water mobility and in conjunction with remotely operated vehicles, and offered a far greater flexibility to diving operations – in much the same way as manned and robotic EVA systems would expand the capability of astronauts in space or during surface explorations.[5]

**Advantages and disadvantages of atmospheric diving suits**
Although developed for underwater diving, atmospheric diving suits offered a technological learning curve for the development of hard pressure garments for spaceflight operations.

The 'man versus robotic' argument has been voiced in undersea exploration as well as space exploration. Both are alien environments in which any living being is at constant risk. In their paper 'Atmospheric Diving Suits', Thornton, Randell and Albaugh addressed the question 'If we can do it with robots, why risk human life?' by observing: 'It is not always that simple a question. ROVs [remotely operated vehicle or vessel] have certainly surpassed atmospheric diving suits in their ability to go deeper. And, despite a nearly flawless safety record, the inherent risk with ADS is a human one – something that you do not have with ROVs. There is, of course, a certain risk any time a diver enters the water, and the human body was not meant to be subjected to extreme pressures, but with the proper redundancy and safety features built in, it can be argued that the risk of serious injury is greater when you drive your car to work every morning'. A similar case can be argued for space exploration.

The authors of the report also reasoned that it was clear that ROVs could perform many of the same tasks assigned to a human diver and to far greater depths, but this was not a guarantee of being faster, cost effective or totally successful. It is a balance of technology, time and finance. Space exploration has not yet developed into a commercial endeavour, but several popular science fiction films have featured commercial operations in the Solar System, including mineral extraction on asteroids, moons and planets. On Earth, mining for oil, coal, gas, precious metals and gemstones has been carried out for centuries, and it is logical that we will move into space to continue this work to support our population expansion, but with the added burden of working in pressure suits supported by advanced remote robotic techniques.

Thinking of the commercial and financial applications of future operations off Earth, the writers stated that the status of today's oil and gas industry could easily be applied to a future mining operation in space – perhaps on Mars: 'The offshore industry means money, and usually lots of it. As the price of oil hovers around $40 per barrel, with often many thousands of barrels per hour at stake until a repair job is completed, hundred of thousands of dollars are at risk.' Operators of deep-sea atmospheric diving suits also realised that the ADS, regardless of its complexity, would never totally replace the ROV or diver. For spaceflight operations, unmanned spacecraft, automated systems and astronauts must work together to produce the best possible return on investments. Dan Kerns – manager of the Diving Division of Hardsuit Inc – has stated: 'The ADS is not meant to replace the ROV, diver or *vice versa*. It is just another tool to put in your bag and pull out when the logistics, cost analysis and so on proves that that tool is the one to use.' This observation will be the same when astronauts explore the surface of Mars and need the best tools available for the assigned task.

The advantages of using ADS include the removal or reduction of the dangers of physiological hazards while working at ocean depths, and the ability to place a human operator directly at the worksite, where the skills of human improvisation and decision-making are paramount. An operator in an ADS has a greater range of dexterity than the ROV, and depth perception far greater than visual observations via the remote vehicle. With no long period of decompression or compression, a diver can also translate to and from a worksite much faster than a scuba-suited diver, and to greater depths, offering greater flexibility for a working envelope. The ability to carry additional and special tooling on the suit also offers greater variations in working patterns than does the standard hand-held tool-kit of scuba divers, and the ADS eliminates the need for decompression chambers – especially in locations where access is limited, impractical or costly. The use of an ADS system greatly extends the operational working period from minutes to many hours.

Its disadvantages include less payload capability than a total ROV, and it has limited access to confined locations. Most importantly, although life support is less than a saturation system, it is still more than with a totally ROV system, so human safety and emergency and contingency systems must be included to support life and to care for injured divers when and where necessary.

These factors have direct application for extended EVA operations planned for Mars.

**From sea to space**
Thornton *et al.* defined the atmospheric diving suit as follows: 'An articulated anthropomorphic single-person submersible capable of diving to depths of up to 758 m while maintaining the internal pressure at or very nearly 1 atmosphere. The immediate and obvious advantage of the ADS is its elimination of most of the physiological hazards of ambient pressure to the diver. There is no need for compression or decompression schedules, no requirement for special gas mixtures, and no danger of nitrogen narcoses or the bends. Likewise, the ADS can venture to and down the water column any number of times without any consequences or delay.' With in excess of 200 years development of ADS technology, the primary objective remains the same as with less than 100 years of altitude and space pressure garment development: supporting a human operator in a hazardous environment while aiding in the execution of assigned tasks.

In 1967 the US Litton Industries Space Science Laboratories revealed plans to develop a new ADS suit, capable of operation to 182 metres, and featuring a combination of constant-volume convolute joints and rotary joints, duplicating as closely as possible the wearer's joint articulation. Called the Underwater Experiments (UX-1) suit, it represented a cross-over between research into stratospheric, terrestrial and space pressure garments and those developed for underwater explorations, and incorporated the experience of the company's work on industrial and atmospheric suits, applying them to underwater research. It was far in advance of anything that had gone before, but it never entered production.

**Ultimate high-altitude pressure suits and cabins**
While developments in underwater suit technology occupied the naval and oceanographic science world, so the development of atmospheric high-altitude pressure garments became the focus of air forces and atmospheric scientists in the early decades of the twentieth century. After centuries of dreaming of flying into the air and eventually to the stars, the advent of the balloon finally made human flight possible in the eighteenth century; but for more than 150 years these aeronautical

adventures were limited to heights of a few thousand feet, due to the thin atmosphere at higher altitudes. As with underwater exploration, entry into the upper atmospheric layers plays havoc with the human body. The human form is designed for optimum operation at sea level, and taking it into environments in which it was not designed to function, without appropriate protection, is potentially lethal. Apart from the obvious lack of oxygen there are extremes of cold, friction caused by flying creates heat, and protruding from or leaving the aircraft at a speed exposes the aeronaut to high-speed winds.

As early as 1901, German balloonists, with the support of an oxygen supply, had reached 35,424 feet. By the First World War, as the operational height and speed of aircraft increased, methods of crew protection were required, and the aircraft engine supercharger, supplying air to the cockpit, was developed. During the 1920s, pilots flew to altitudes of between 30,000 and 40,000 feet, and survived – just. Flights above 40,000 feet would require additional oxygen (though in 1921 a Lieut Macready attained 40,800 feet). The intense cold, the lack of adequate protective clothing and the rudimentary oxygen supply equipment almost killed the pilots, and tests in altitude chambers revealed that at altitudes greater than 45,000 feet the body could not take in enough oxygen to survive unless an external pressure level was artificially maintained. One option was the sealed and pressurised cabin, which began to be developed during the 1920s.[6]

From the 1930s, crew-protection pressure garments were designed for internal high-altitude applications, until more reliable pressurised cabins became available. However, over the next thirty years the development of both a personal pressure garment and a pressurised crew compartment provided the first space explorers with adequate protection in the short pioneering journeys around the Earth and for the first excursions outside the spacecraft while developing EVA techniques in the vacuum of space.[7] But with the desire to send men to the Moon and eventually on to Mars, a new type of suit would be required to allow more mobility during surface exploration.

## SURFACE EXPLORATION SUIT DEVELOPMENT

'Like the spaceship itself, a spacesuit must be designed for the specific job to which it is assigned. For example, just as a spaceship designed to affect a landing on a body with an atmosphere will differ from one which had to alight on an airless body, so the spacesuit which is to do duty on the Moon will differ in detail from one that has to operate in the presence of an irrespirable atmosphere.' So wrote H.R. Ross in a detailed article on the design of a future lunar spacesuit[8] – one of the earliest serious design studies for a pressure garment to be used during surface exploration of the Moon.

Noting the primary considerations that the Moon, according to the evidence at that time (the late 1940s), lacked an appreciable atmosphere and underwent great extremes of surface temperature (quoted as between 120° C (248° F) and −150° C (−238° F)), the paper highlighted the need for a self-contained oxygen supply and a system for removing carbon dioxide and excess water vapour, either as part of the suit or as 'an attendant vehicle'. In discussing the provision of suitable breathing systems, Ross referred to the work conducted in previous 'breathing kit' studies and experiments concerning diving, mountain-climbing, high-altitude flying, mining, and general work in 'irrespirable atmospheres'. Referring to the work by Sir Robert H. Davis in his books *Breathing in Irrespirable Atmospheres* and *Deep Diving and Submarine Operations*, Ross also recognised that a fully self-contained system offered 'the greatest freedom of action.' The British Interplanetary Society study also addressed other questions concerning the design of a lunar surface exploration suit, such as internal temperature control, boot insulation for direct contact with the surface, helmet design, mobility, viewing ports and eye protection, gloves, communications, sizing, and radiation protection. More than a decade later these same considerations were addressed during the development of the Apollo lunar suit.

### Suited for Apollo

Due to the Apollo mission profile and the size limitations of the spacecraft, the internal transfer tunnel and the EVA hatches, the suit design chosen for Apollo also

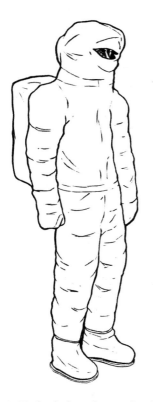 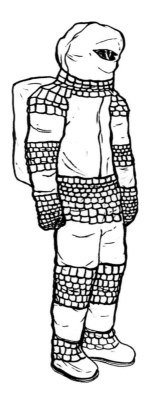

Early designs for an Apollo spacesuit with added thermal coverings for exploring the Moon.

needed to provide personal protection in the event of cabin depressurisation, launch, entry and docking, as well as providing mobility and protection on the Moon throughout a mission of 7–14 days. For the first time in the space programme, the Apollo spacecraft afforded a reasonable amount of internal volume for the crew to take off their suits and put them on again as required during the course of their mission. As only two of the three astronauts of an Apollo crew would explore the Moon, the third crew-member did not require the additional features incorporated in the two lunar explorers' suits.

The development of a lunar surface suit for Apollo had evolved from the early 1960s. It was based on a pneumatic pressure garment system, and was in some respects governed by the limited dimensions of the spacecraft. Various designs and configurations were evaluated over several years, but the final version had to have far more mobility and be comfortable, functional and safe; it had to record metabolic rates and biomedical data, support hygiene requirements, supply emergency and contingency facilities, and be capable of utilising a wide variety of support equipment to deploy experiments, retrieve samples, take photographs, and carry out observations, as well as communicate with each other, the crew-member in the orbiting CSM, and Mission Control on Earth. It also needed to incorporate additional means of protection against extremes of temperature, low gravity ($\frac{1}{6}$ that of Earth), primary and secondary micrometeoroids, and increased visual hazards.

The resultant Apollo lunar surface pressure garment is not the primary subject for this book. (It is described in detail in other works listed in the Bibliography). In summary, however, the EVA unit (A7L for Apollo 11–14) consisted of a Nomex comfort layer, a neoprene-coated nylon pressure bladder, and a nylon restraint layer. The outer EVA layers were fabricated from an integral thermal/antimicrometeoroid cover consisting of a liner of two layers of neoprene-coated nylon, seven layers of Beta/Kapton spacer laminate, and an outer layer of Teflon-coated Beta fabric.

Next to their skin the astronauts also wore a liquid-cooling nylon–spandex garment consisting of a network of plastic tubing through which coolant (water) from the Portable Life Support System (PLSS) was circulated. On their backs they wore a PLSS back-pack assembly supplying, on the early landing missions, 4 hours of oxygen at 3.9 psi, coolant (water), and a subsystem for handling the returning oxygen cleansed of solid and gas contaminants by a lithium hydroxide canister. The unit also contained a communications system, a telemetry system, displays and control units, and a main power supply system. A thermal jacket was installed over the unit, and on top was mounted the oxygen purge system that provided a 30-minute contingency from two 2-lb bottles. On their heads they wore a 'snoopy' communications cap and an inner pressure bubble helmet, over which was a polycarbonate shell, with two visors incorporating thermal and optical coatings to provide the wearer with protection against impacts, micrometeoroids, and thermal, ultraviolet and infrared radiation during EVA.

The lunar EVA gloves were made from an outer shell of Chromel-R fabric and thermal insulation layers to provide additional protection when handling hot and cold objects in a vacuum. Fingertips were protected by silicone rubber caps on each digit, and lunar boots worn over the inner garment and boot assembly provided toes,

An early design (1960s) for a semi-rigid spacesuit.

sole and heels with protection against abrasion and extremes of temperature. The boot liner was composed of several layers to both protect the astronaut's foot and aid walking on the lunar surface, and was made of Teflon-coated Beta cloth. Other layers were fabricated from aluminised Mylar, Nomex felt, Dacron, Beta cloth, and Beta marquisette Kapton. The outer layer was a shell assembly of Chromel R metal fabric. The sole of the boot was made of high-strength silicone rubber, with a moulded tread layer for grip.

Small upgrades were incorporated after each flight, as experience and feedback was gathered from the early landing crews; but a significant improvement was already in the planning stages for the later J-series missions, resulting in the A-7LB designation. These new suits included additional mobility at the waist and neck, with relocation of entry zips from the crotch allowing more leg mobility. With the new suits it was also easier to stoop on the surface, and there was less shoulder pressure on the wearer as well as improved thermal garments and larger in-suit drink bags. On later missions the PLSS was improved to increase the water supply, and there was a larger power battery and additional lithium hydroxide facilities. Due to the improvements, surface EVAs could be increased to around 7 hours.

The combination of suit, helmet, gloves, boots and back-pack – weighing approximately 300 kg – was termed Extravehicular Mobility Unit (EMU). The system allowed for up to two surface EVAs on Apollos 11–14 and up to three on Apollos 15–17, with contingency for an in-flight EVA transfer from the LM to the CM should docking-tunnel transfer not be possible.

### Suited for the Moon

Surface operations using the Apollo pressure garments totalled in excess of 80 hours (160 man-hours) in fourteen excursions between July 1969 and December 1972. The reliability of the pressure garment was one of the success stories of Apollo, allowing twelve men to explore the Moon in a series of surface activities that were an outstanding success and were completed without any serious injury to any crew-member. After years of theoretical studies and more than a decade of direct development, the Apollo lunar surface garment remains our only extraterrestrial surface exploration suit with operational experience. Although other designs in both the US and the Soviet Union were evaluated, only the A7L/A7LB finally went to the Moon. Despite this success, advances in technology and a change in objectives and hardware will ensure that the next lunar suit will not simply be an upgraded version of the Apollo suit, and nor will the suit for Marswalks.

The astronauts generally reported that use of the suit on the Moon was satisfactory. According to the Apollo Program Summary Report, moving in the $\frac{1}{6}$ gravity while wearing the Apollo EMU was not difficult, and adaptation was quite natural. The astronauts soon found that a two-footed loping gait or a running gait worked very well, but the mass of the back-pack and the apparent slipperiness of the

A design for an advanced Apollo lunar surface exploration suit.

lunar regolith required a substantial stopping distance. No significant discomfort was reported (due to sufficient heat removal even in the high temperatures of the lunar day), even after some degradation due to a build-up of lunar dust. The Apollo 11 astronauts were the first to notice the sharp contrasts of sunlight and shadow (which it had not been possible to simulate on Earth), but it took them only a few seconds to adjust to the new lighting conditions, and had no detrimental effect on their work performance. Judging distances remained a challenge, and subsequent crews were helped by the addition of sunshades on the helmet visor assembly. They also supported the need for improvements to the suit to facilitate easier bending and kneeling. The one memorable feature of wearing the suit was the hand-fatigue and tiring of the arms caused by carrying ALSEP packages. The design of the Apollo suit included a glove pressure that was difficult to maintain over sustained periods. Hand strength alone was not always sufficient to grab, hold or continually manipulate objects.[9]

It was clear that for the two or three days on the Moon, the A7L suit system was satisfactory for up to three short excursions of up to 8 hours; but for more extensive surface operations on the Moon, and on Mars, other pressure-suit systems would need to be developed, with more operational flexibility over longer periods. Even as Apollo was being developed, work was underway on a programme of 'advanced EVA suits' to follow those early lunar landings. This work – originally planned for post-Apollo operations on the Moon and in Earth orbit – has continued over the past three decades, and is now the basis for planning more definitive pressure garments for the human exploration of Mars.

## EARLY HARD SUITS FOR SPACE

'As future spaceprobes and advanced missions for extended orbital, lunar and planetary adventures are defined, more complex and unique space suits will be called for,' wrote C.C. Lutz, in a 1966 paper detailing the development of pressure suits and support equipment up to 1965.[10] In his summary, Lutz suggested that until an 'ideal suit' could be developed, wearers would have to 'put up' with the inherent disadvantages of wearing 'conventional' suits.

This 'ideal suit' was thought to be 'one that would come in the form of an aerosol spray' that could be instantly 'applied' when needed. It would occupy no space, impose no discomfort and no restrictions, present no interface problems and be as easy to use as hair spray. The package and method of dispensing a suit of this concept is available. The only detail that remains to be 'worked out' is what to put in the can. While this, at present, seems beyond the realm of serious possibility, it stresses the belief that 'future protection and work in space must be supported by unique approaches in design and operation.' Almost forty years later, the concept of what is also commonly referred to as a 'skin-suit' remains for future generations, and although it is still being investigated it will probably remain only in science fiction (at least for now). Lutz also considered that spacecraft systems would evolve to reliably sustain crews in shirtsleeve environments, and that special lightweight emergency intravehicular pressure garments would be available to minimise the burden of wearing full pressure garments for the duration of the mission. Future suits would 'undoubtedly be mission orientated.'

The paper addressed future lunar exploration after Apollo, during which extensive EVAs and surface operations would be mounted from an operating base. In doing so, new approaches to the design of space suits would have to be addressed. These objectives of long-term reliability, improved mobility, low suit-leakage, high durability against repeated wear, and lightweight construction, though considered here for lunar and space station operations, can equally be applied for future planetary suit design out at Mars. At the time that the paper was written, work was already being conducted on a series of so called 'hard' suits to meet these goals.

Prototype hard suit technology (with metal and plastic) was already demonstrating very low rates of leakage, great durability, and reduced vulnerability to abrasion and puncture, compared with conventional suits. They could operate at pressures of around 7 psi, resulting in low operating forces on the wearer and good shoulder and arm mobility. The major problem with such a suit was the volume of storage space it required when not in use.

### The RX suit programme

The first extended evaluations of hard suit technology for space operations came under the programme of studies carried out by Litton Industries and NASA in the 1960s. These studies were focused on advanced lunar spacesuit designs, but have provided a focal point for all advanced hard suits that have followed. In his 2001 study on the origins and development of Advanced Suit Technology, author Gary L. Harris provides a detailed account of the original design and development of a complicated series of suits that were inspired by work conducted by Litton Industries in the 1950s.

Work on hard suits began at Litton under a programme connected to thermionic valves used in the electronics industry during the 1950s. Although linked to aspects of a future missile and space programme, the problems in developing this technology necessitated the radical approach of placing a suited technician inside a vacuum chamber to conduct work on the valve and associated equipment while the chamber was operated at a vacuum. Studies of available suit technology at that time revealed that a new suit would need to be constructed to meet the requirements, and so the Litton Mark 1 suit was conceived. It was a hybrid design of an aluminium alloy hard upper torso with an integrated helmet and cannibalised leggings from an old USN high-altitude pressure suit, as leg mobility was not a major objective for the suit's use in a chamber. The bellow-type joints caused a variety of problems, and a system of rolling convolute joints was therefore incorporated in the arms. These consisted of a series of bracing rings that allowed the fabric of the arm to telescope for a two-axis movement that also retained internal volume when in use. In the late 1950s, examinations and tests of this suit by Litton staff and USAF doctors and test pilots (including X-2 pilot Ivan Kincheloe) resulted in a development programme of evaluating hard suit concepts for advanced lunar exploration. In 1958 – when NASA was formed, and three years before President Kennedy's commitment to the Moon –

Designs for an advanced Apollo suit undergoing simulated EVA tests in the 1960s.

the USAF generally referred to the pioneering Litton suit as the 'USAF Mark-1 Model Extra Vehicular and Lunar Surface Suit'.[11]

The suits that followed in this programme through to 1969 were designed to evaluate a lunar suit, and are summarised here.[12]

*RX-1* This was an early effort to respond to President Kennedy's challenge, at a time when most of the technology required to achieve the feat of allowing a man to walk on the Moon did not exist. As Harris observed in his book, there was no point in going to the Moon if an astronaut could not walk on the surface and perform useful work. The availability of the Litton chamber allowed early studies to be completed on what was eventually termed the Rigid EXperimental (RX) pressure suit 1. It was estimated that such a suit would provide a lower leakage rate, more resistance to wear, and a far greater protection against micrometeoroids. RX-1 was delivered to NASA in the early months of 1964, and served as a test-bed for future developments of this type of suit. In tests, it proved to have excellent mobility in the arms, shoulders and legs, due to its rolling convoluted joints, allowing a complicated series of movements while retaining constant volume and control. Zippers were not used to close the suit; instead, there was a device resembling a submarine hatch locking device, termed a 'dog lock'.[13] Setbacks included the overall weight of the unit (37 kg – 82 lb), and the design of the hip area, which proved restrictive in mobility. The operating pressure was 35 kPa (5 psi).

*RX-2* Delivered in December 1964, this design featured programme objectives of developing a constant-volume hip-joint and evaluation of the possibility of waist mobility based on experience with the RX-1. Milestones in its development included a low-torque constant-volume hip joint. Problem areas included excessive weight (40 kg – 88 lb), and a tendency to be 'top-heavy and unstable' due to its high centre of gravity. It was also found to be restrictive in sizing adjustments and was difficult to don.

*RX-2A* This revised version weighed 4.5 kg (10 lb) less than earlier models, and had reduced shoulder width. Objectives for this suit included a self-donning capability, additional protection against micrometeoroids, improved joint systems, improved waist mobility, and 'compatibility with Apollo mission environmental guidelines'. Delivered in June 1965, the suit's milestones included improved methods of donning and doffing, although a lack of complete sizing and total self-donning capabilities still caused problems. Another difficult area was in reducing overall weight, but a two-plane body seal closure and a composite aluminium and fibreglass honeycomb sandwich structure was developed, together with a three-axis shoulder (armoured convolute) section, an improved waist locking joint, a hemispherical dome helmet for improved vision, and a self-supporting suit structure. This unit also demonstrated excellent ankle mobility.

The 'shoes' of the RX series were designed along the lines of the Dutch sabot. Litton engineers purchased a pair of these shoes for less than $5 to evaluate the most useful contour for lunar boot sole and heel design. Using the Dutch shoe (minus a shaved-off toe) for a mould, a sole insulation of polyurethane was added inside to create the final design. This was an excellent example of conservatism in developing

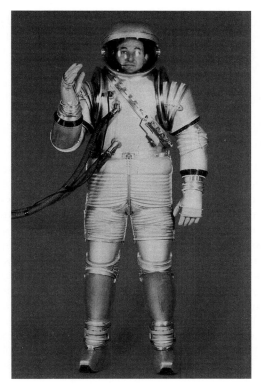

A prototype hard suit.

space technology. The funding for RX-2A in Fiscal Year 1964 was $437,000, but only $5 of this was spent on sourcing an adequate pattern for the boot design. Plans to include a test version of the suit in early Apollo missions proved difficult due to weight and bulk, but it became a valuable ground-test article, being used on the Apollo 'rock pile' at MSC and proving to be rugged and reliable for surface exploration. Harris alluded to this in comparisons with the chosen Apollo soft suit design: 'The RX-2A probably underwent more testing than any other hard suit to date. It proved itself rugged, reliable, and forgiving of neglect in maintenance. Years later, many lunar astronauts complained of the Apollo soft suits being nearly worn out by the time their short missions were complete. It seems obvious that a hard suit could have alleviated some of these wear problems, if a means of stowage had been found.'[14] This also demonstrated the potential for hard suit application in post-Apollo and possible planetary surface programmes.

*RX-3* Delivered to NASA during 1966, this suit was designed to provide a 10–90% sizing capability, improved mobility in the shoulder, easier kneeling capability, direct interface with the Apollo Portable Life Support System (PLSS) and Liquid Cooled Garment (LCG), improved internal comfort for the wearer, and a reduction in weight (28.5 kg – 63 lb) and overall bulk. Its milestones included the development of completely self-donning and doffing capability, with an improved sizing range and overall improved mobility due to a lower centre of gravity. Using the suit as a 'simulator', the test programme demonstrated the wearer's ability to climb in and out of the Apollo LM, and the ability to interface with the PLSS and LCG. Despite using magnesium and magnesium–lithium alloys in the rotary seals and shell fabrication, there were still problems with the excessive weight and bulk of the design; but the tests also included the development of a compact integral back-pack that employed heat exchangers instead of liquid cooling which, it was hoped, would reduce the donning time.

*RX-4* Developed in 1966–67, this version was designed to address the associated suit technologies of passive thermal control, hard suit glove development, intra-suit connectors, a thermal boot, and prototype EVA visor assemblies that were directly compatible for use on Apollo suits. The development programme identified problem areas, such as the continual reduction and holding of the upper weight limit within the guidelines, and a strong need for mission definition. Spin-off applications of this suit included adapting its constant volume wrist joint and hard suit glove into the 1-atmosphere arm–glove chamber system in the Lunar Receiving Laboratory, used for handling returned lunar samples. A development of the thermal boot was used on the Apollo 4H (Hamilton) suit as part of an in-house evaluation to determine its feasibility for use in the Apollo lunar programme. The visor was also evaluated for use on Apollo suits for protection against infrared and ultraviolet radiation. The mass of the suit remained a constant problem, but it was addressed by Litton by 'nesting' the suit to fit in as small an area as possible (now a feature of the current Shuttle EMU).

*RX-5* This was the final development of this series up to 1969. It was 6–22 kg (13–48 lbs) lighter than the RX4, and had increased mobility to ease twisting, bending and

Mobility and reach tests of an advanced suit for planetary exploration.

even touching the toes without reducing the internal volume of the suit. For operational applications it featured standardised compatibility with the equipment used in the Gemini and Apollo programmes

Although developed in support of advanced lunar exploration suits, these models provided baseline data for adapting hard suit technology to spaceflight operations, and in turn applying its spin-offs to deep-sea diving programmes – in which this type of research had begun more than a century earlier.

The Litton Company published its own report on the results of the RX programme, including the RX-5. 'When utilised with mixed gas atmospheres and operating at 1 atmosphere, mobility of the joints should not pose serious problems for joint manipulation and torque.' The RX-5 development resulted in a suit able to operate at 7.5 psi, with the weight of individual components limited by the state of current (1969) technology and manufacturing techniques. The introduction of inert gases, such as helium or nitrogen, was not found to complicate the design of the life support system, due to low leakage rates. Should the leakage rate increase, however, the internal pressure could be quickly lowered to 5 psi, and the pure oxygen supply reintroduced into the LSS to compensate. These studies with helium/oxygen systems were key factors in the evolution of subsequent spacesuits and EVAs, in which time, mass, safety, and the ever-present cost factor, had to be incorporated.

**Suited for Space Shuttle and space station**
At the same time that Litton was evaluating its range of RX suits for NASA, the agency's own Ames Research Center was also evaluating a series of hard suits (designated AX – Ames EXperimental) in cooperation with the AiResearch Division of the Garrett Corporation. For these suits, multiple-bearing technology was studied and developed. Litton continued its own work on advanced EVA suits, which were generally the forerunners of the suits developed for Shuttle and space station operations, and not true surface-exploration suits. However, the experience and operational data gained from using the Shuttle EMU over twenty years, plus the development of suits for use on the USAF Manned Orbiting Laboratory (which was later cancelled), Skylab, Freedom and the ISS, has provided a valuable database for supporting EVAs from a space station (and to some degree, from a deep space vehicle).

Both Litton and AiResearch embarked on Advanced EVA Suit (AES) programmes. These were not fully 'hard suits', but a hybrid combination of elements of hard and soft suits. They were initially planned for use on later Apollo flights, but the lunar programme was curtailed and eventually abandoned, and the work was redirected towards the Apollo Applications Program (space station), with the goal of using the increased mobility, protection and reliability of hard suits, coupled with softer elements for ease of storage in restricted-volume spacecraft or space stations. The launching of compact pressure garments for constant and extended use from a space base has been the outgrowth of these programmes over the past three decades. (A clear explanation of the development of the suits that evolved into Shuttle and Freedom studies is provided by Harris.[15]) Although aimed more at Earth orbital operations, some of the technology developed for these

An advanced Shuttle/space station EVA suit with rear hatch entry and integrated helmet.

pressure garments has been evaluated for 'planetary surface suits', alongside the fully 'hard' suit pioneered by Litton.

### Soviet developments

During the 1960s, the Soviets developed their own spacesuits from a pressure garment programme that had evolved over the previous thirty years and had scientific, medical and military origins. Unlike the chosen design of the Apollo pressure garment, the Soviet lunar suit Krechet featured a hard torso, soft leg and arm elements, and an integrated helmet. Entry was via a hinged hatch in the integrated life support system back-pack, and though the suit was not used on the Moon (nor, indeed, in space) it was a useful development tool. An orbital-only version of the lunar suit – Orlan – provided the basis, from 1977, for the series of space station EVA suits used on Salyut 6, Salyut 7 and Mir, and continues to be a valuable operational unit in the construction and maintenance of the ISS.[16]

Cooperation between Russia, the European Space Agency and NASA has resulted in several programmes for improving techniques and hardware for EVA operations, both at the ISS and on future programmes. This has focused research into providing suitable pressure garments for walks across the surface of Mars.

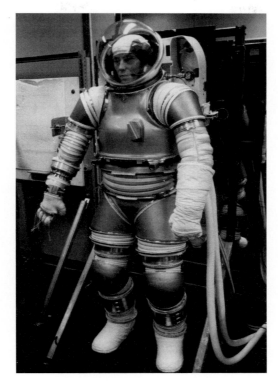

Prototype of an advanced spacesuit undergoing 1-g tests.

## THE FIRST PLANETARY EVA SUIT

The design of the first pressure garment for operational use on a planet other than our own has yet to be defined. Pressure garments have been designed and produced for more than seventy years, and they have been used in the exploration of space (including on the Moon) for more than forty years. From those experiences, numerous pressure garments have been designed to evaluate both the return to the Moon and the exploration of Mars. None of these are operational suits, but they have proved very useful test and demonstration models along the road to defining just what the first martian EVA suit will be required to do to support the first excursions across the surface.

### Physical and environmental limitations
Designers of pressure garments for use on Mars must consider certain physical and environmental limitations. A productive timeline will be limited by assembly, maintenance and operational overheads, exercise, sleep, meals, and communication coverage; and mobility will be limited by the type of transportation, the mass of the suit, its consumables and tools available and carried, and the degradation of the five

Microgravity evaluation, in the KC-135 aircraft, of a rear entry pressure suit.

human senses, which can only be supplemented while wearing a fully enclosed pressure garment incorporating information aids and sensors. The environmental considerations include the following:

- The risk of exposure to radiation.
- Extremes of heat and cold, which vary with altitude, the seasons, and the day and night cycles.
- Pressure: a 0.001-atmosphere carbon dioxide-rich atmosphere requires the development of special carbon dioxide and thermal subsystems.
- Lighting conditions that constrain the effective work-time and the distance that the astronauts could travel in unfamiliar regions, either with or without artificial light sources and power systems.
- The ever-present dust, which will degrade pressure seals, obscure vision, and adhere to hardware, equipment, experiments and solar-power arrays.
- The winds, which will exacerbate erosion of materials and suit components by dust particles, obscure hardware, and possibly move unsecured hardware.
- Extended exposure to microgravity and $\frac{1}{3}$ g, which will have a deteriorating effect on bones and muscles.
- Two-way organic contamination, which will impede productive work.
- The roughness of the terrain and the instability of surface material, which will impede site access as crews tackle slopes, cliffs, and other obstacles.
- Material hardness.

**Information technology**

More needs to be known about the martian environment before defined suit development can commence, and improved technology is required to ensure efficient working in safe conditions. According to NASA EVA office presentations in 2000, environmental data are lacking in the following areas:

- Levels of ultraviolet and particle radiation at the surface.
- Seasonal, daily and altitude variations of atmospheric composition affecting temperature, pressure levels, dust, natural lighting, and the speed and direction of the winds.
- Impact of dust particles and winds in convective/radiation heat transfer and solar flux.
- Chemical composition and reactivity of the soil/dust.
- Electrostatic charge.
- Soil mechanics: bearing strength, penetration, resistance cohesion, adhesion and abrasion.
- Levels of trapped pressurised fluids and gases in the atmosphere, and volatile gases and toxic materials.
- Touch temperatures of surface and subsurface materials.
- Short-term and long-term effects of corrosion and abrasion of suit materials and coatings.
- Terrain characteristics and location maps, including slopes, cliffs, caves, ravines, craters, obstacle size and distribution, instability of the surface, and hardness of the subsurface and rock strata.

The technology still required includes:

- A lightweight, portable and reliable life support system.
- A range of defined surface transportation modes, airlocks, and information and navigational aids.
- A suite of robotic and support facilities.
- Effective radiation protection.
- Capability in gathering, converting and using *in situ* resources.
- Compact power sources.
- Facilities for care and storage of gathered samples.

While it is true that there still remains a significant amount of work to be completed before a definitive suit design and EVA programme is manifested, work has already begun on narrowing the developmental focus of a potential Mars EVA suit and support hardware.

**New ideas for a new goal**

In the late 1980s the Russian company Zvezda began feasibility studies into the requirements for a safe and functional planetary pressure suit. These studies were conducted under the Mars Exploration Project, which the Soviet government's science and technical programme listed as a future possibility. Using its considerable experience in developing pressure suits for the national space programme, Zvezda wanted to evaluate possible obstacles that would need to be eliminated.

Zvezda already had experience with the development of the Krechet suits for the (later abandoned) Soviet manned lunar landing programme, and was also the primary contractor for the Orlan series of pressure garments used by Russian cosmonauts on Earth-orbit EVAs since 1977. Orlan suits have been used during more than twenty-five years of operations; but they were designed to be used only in open space and have limited mobility, and would therefore need to be significantly modified for use on Mars. Experience gained by the Americans during the Apollo missions revealed that the lunar suit needed to be more mobile in the waist, hip, knee and ankle joints to enable the wearer to more easily squat, bend and stoop, exit and enter the Lunar Module, climb onto and alight from the rover, and deploy experiments, gather samples and erect equipment. The Orlan suits have minimal mobility in these areas, and are not suitable for use other than in free space. Similarly, the Shuttle EMU suit system – in operational use since 1982 – was designed solely for orbital operations from the Shuttle (or space station), and not for surface exploration.

How, therefore, can the ideal suit design for a Mars mission be defined? In reviewing both the Apollo surface suit and those used with great success over a long period of time, it became clear that perhaps the best way forward was to design a suit that could be effectively used both during long interplanetary coasts and on the surface of Mars, with a proven long-duration lifetime and the capability of system replacement and consumable regeneration without occupying too much crew-time. In selecting the final design, several parameters need to be considered:[17]

*Compatibility* Any design chosen for a Mars suit would have to be compatible with both the objectives of the mission and with all the hardware chosen to complete the mission profile. It would also have to be compatible with EVA support facilities and hardware (airlocks, rovers, restraints) used in the trans-Mars, Mars orbit and trans-Earth trajectories, as well as on the surface. It might also be possible to use interchangeable parts to allow for differences in the size of crew-members, or for the task assigned, such as replacement leg components to accommodate less leg mobility in interplanetary space and more leg mobility on the surface.

*Durability* EVAs on Mars will be much more frequent than the EVAs on space stations, and in far greater number than during the lunar surface operations of Apollo. The suit would need to support the planned EVA programme on the surface of the planet, both in the number of crew-members involved in each EVA and in the duration and frequency of those EVAs. The suits must also support EVA operations during trans-Mars and trans-Earth coasts and orbital operations – possibly also including sorties to Phobos and Deimos. With built-in contingency and back-up options, the suit and its components would need to be durable, capable of regenerating its consumables, and fully or partially replaceable or repairable on the main spacecraft, in the landers and, in some instances, in a pressurised rover or during surface operations in an emergency.

*Mobility* The design would need to be compatible with the types of task expected to be performed during surface EVAs. In addition, a mobile vehicle (pressurised or unpressurised) would need to include component and system compatibility (with potential recharge capability) with the suits and life support systems, and the suits would need to be sufficiently mobile and flexible to enable the wearer to attach and detach connections or equipment without jeopardising suit integrity. The additional encumbrance on Mars is the presence of dust, as well as localised winds and temperature variations, that could reduce mobility and visibility and affect suit operations.

*Mass* A suit that functions both in deep space and on the surface would not only reduce the number of suits that would need to be carried but would also reduce the overall mass of the suit inventory and the volume required for storage. A reduction in the mass of individual suits would also be important to cope with the $\frac{1}{3}$ gravity. The mass of an Orlan suit is 112 kg – which is more ideal for operations in open space, as the higher mass allows more stability while moving across the surface of a space station. The Shuttle EMU mass is even higher; but it has been estimated that on Mars the optimum mass for an EVA suit will be about 55 kg. Further studies are still to be undertaken before a definitive design is selected; but this will inevitably change, and the incorporation of new materials will reduce weight. New subsystems and operating procedures will also help to minimise the overall mass.

*Environment* As well as providing a suitable internal atmosphere (oxygen, heating, cooling, and emergency contingencies) and a suitable system of water collection and nutrient provision during long EVAs, the Mars suit will also need to maintain a safe and comfortable environment while protecting the wearer against the vacuum of space, radiation, atmospheric and soil conditions, and material and system wear and deterioration.

*Thermal properties* The suit will need to operate under extreme low-pressure and temperatures, and at the same time sustain the wearer, with contingency built in. To dissipate heat when temperatures increase, and to retain or supply heat when temperatures drop, it will need to include heat sources, exchangers, and thermal coverings. The systems used in orbital suits (heat exchanger/sub-laminator) are not suitable for use on the surface of Mars, due to the planet's low atmospheric pressure (7–10 HPa). During the Apollo missions, the conditions on the Moon caused particles of dark material to adhere to the surface of the suit, effectively blocking its white outer covering and affecting its heat-reflectant properties, so that its coolant systems had to work harder. On the Moon, the problem of adhesion was exacerbated by the combination of the vacuum, the low gravity, the static charge caused by exposure to the ionised solar wind, and irregular-shaped particles. On Mars, however, with an increased absorption of the thermal infrared radiation of the soil, adhesion is not expected to significantly raise the temperature of the suit, and in some cases the thermal coverings may intentionally be produced in darker colours.[18]

*Safety* All elements of the suit will need to be man-rated – free from sharp edges, protrusions and other hazards that might affect its operational safety. Procedures and contingencies will also need to be built into the suit and its operation to allow 'rescue' of crew-members in difficulties (such as the 'Buddy' system of sharing life support subsystems in the event of a failure of the Apollo lunar suit during surface exploration). All EVAs will also have to be conducted under strict guidelines, with a walk-back capability available to a safe haven (shelter/pressurised rover/lander) in the event of a suit system failure requiring immediate termination of the EVA. EVAs will certainly be conducted in teams of astronauts, providing redundancies and a 'buddy' system for completing assigned tasks, while monitoring suit parameters and local weather and surface conditions and remaining in contact with colleagues and the EVA choreographer/Capcom or lander mission control. Orbital monitoring of surface activities for emergency situations will probably be too remote for adequate protection. Earth-based monitoring will be enabled only in a back-up or analysis mode.

*Reliability* The Apollo suits were used on only short missions of up to 10–12 days, and on the Moon for up to three periods of 8-hour EVAs over three days. Suits for use on Mars will probably be used several times a week, by different crew-members, for varying lengths of EVAs. Their components will have been flight tested, used and maintained in the long flight from Earth, and will need to be available for similar EVA support roles on the long flight home. Such long-life experience has been gained only by the Russians. Although the Shuttle EMU has been used on multiple EVAs, it has been returned to Earth for refurbishment and/or replacement at the end of each Shuttle mission. Even on Russian Salyut, Mir and ISS operations, the additional advantage of the long-duration use of EVA suits is that new units can be manifested as part of a cargo consignment relatively quickly, and can be taken from the launch pad to the station within a few days. This capability will not be possible on a Mars mission, although it is possible that a small amount of equipment could be sent on ahead on unmanned landers, which would need to be accurately landed close to the eventual manned landing site. (Such 'taxi' missions were planned for later Apollo missions, but they were never enacted.)

**Scenarios for EVA suit utilisation**
Studies at Zvezda focused on four scenarios in which a Mars surface EVA suit could be utilised, and also noted the use of robotic vehicles and systems to aid and support EVA activities over long periods, together with rovers (unpressurised with life support regeneration connections and pressurised with airlock capability) for extended surface operations.

The simplest suit would be used only for contingency EVA operations, and would be available in the event of an emergency for transfer from a pressurised rover to the lander. In normal operations an airlock system could be used to aid retransfer from the lander to the rover, which would have full life support capability and would control robotic EVA operations. In a contingency, EVA capability of up to 2–3 hours would allow transfer from the rover to the lander via an airlock in the event of an inoperable primary docking system. If an unpressurised rover were to be used, then facilities to attach the life support system directly to the suit would enable extended EVA operations. Again, moving between the lander and the rover would require autonomous EVA capability of around 3 hours. Both these types of contingency EVA suit could be adapted from existing technology, but they would restrict 'human' EVA activities on the surface.

In the USA, Hamilton Sundstrand has evaluated a planetary EVA suit concept that envisages short EVAs from the rover and periodic changes of back-packs containing defined EVA consumables. It has also been suggested that limited umbilical extensions to suits would allow some local traverse around the rover. However, dragging an umbilical over even a short distance would add to the effort and restrict the distance travelled. As experienced during Apollo, visibility would not necessarily allow an astronaut to be aware of where his feet are in respect to the umbilical.[19] Even with a fully self-contained suit system, EVA surface activities would be possible for only 6–7 hours, with a further 2 hours maximum for contingency and pre- and post-EVA activities. This would be dependent on the capability of the wearer rather than the equipment required for support, because wearing any pressure suit is tiring, dehydrating, stressful and restricting – no matter how the technology develops to alleviate these 'human parameters'. The development of fully self-contained suits will require new technologies and components, and would be dependent upon the final mission design, its objectives and inevitable human parameters, and a defined budget.

**Considerations for access to and from the martian surface**
During Apollo surface activities, the two astronauts helped each other don their suits in the cramped Lunar Module and then proceeded one at a time, on their hands and knees and backwards, out of the forward hatch to the front porch on top of the nine-rung ladder that took them to the forward-strut landing pad and onto the surface. This process had to be reversed to return to the LM cabin for the rest period between each EVA and for donning prior to rejoining the third astronaut in the Command Module in orbit.

One of the most obvious problems with working effectively in a pressure suit is in keeping it functioning and clean. The Apollo missions encountered the problem of lunar dust, and the astronauts who experienced working on the surface and living in the LM often stated that one of the best tools available was the brush for removing the heavier particles of lunar material. Even so, the material invariably found its way into the LM cabin, and the smaller granules ground their way into the gloves, the outer coverings of the suit, and even the pores of the astronauts' skin. A quick comparison of the appearance of an EVA suit in Earth orbit with one on the Moon during the second or third excursion revealed just how much the lunar dust dirtied the suit and effectively blocked out all suit recognition marks. However, it did not cling to coated fabrics or smooth surfaces as much as it adhered to woven fabrics, and it has been suggested that the final design for future planetary exploration suits be fabricated from as many hard and smooth components as possible to help alleviate the dust problem.

The other problem facing future Mars (and lunar) EVA planners is in keeping dust out of the habitable portions of the lander (or base). During Apollo, direct access from the LM cabin to the ladder leading to the surface resulted in dust being carried up into the LM; and even in the pressurised environment it insinuated itself everywhere after the suits were taken off for the rest period. Keeping the habitation areas clean will help enhance the living quarters and protect the environment and health of the crew, who will 'feel' cleaner. There are several ways to enable a crewman to gain access to the surface from the lander:

*Direct access hatch* This Apollo-like system is the easiest to engineer, but on a long surface stay it would be the most troublesome, with the potential of dust material clogging hatch mechanisms, contaminating the internal habitation facilities, polluting the food and hygiene systems, and affecting the eyes, skin and airways of the crew. On Apollo, the problem of dust persisted after the astronauts left the lunar surface, and as early as the second landing – Apollo 12, in November 1969 – Command Module Pilot Dick Gordon insisted that his colleagues undress and clean themselves and their suits, the rock boxes and other equipment before he let them out of the 'dirty' Lunar Module and into his 'clean' Command Module. The return flight on Apollo lasted just three days; but from Mars it would take many months, and dust contamination over that period of time would be a hazard as well as a nuisance.

*Airlocks* These have been used on space stations since the 1970s and the Shuttle since the 1980s, and will probably be incorporated in the final design of the Mars lander. They have proven to be very useful as a location for stowing EVA gear when not in use, and in the ISS programme the Pirs and Quest airlocks allow the crew to maintain and service EVA equipment separately from the main habitation area. One drawback, however, is the human tendency to stow unwanted material in any available space, and the temptation to use the airlock for anything other than surface access must be avoided. A back-up entry system or secondary airlock must also be considered in case of system failure in the primary airlock or outer hatches.

*Curtains and scrubbers* As part of the airlock system, these would include elements of both an air curtain and an electrostatic scrubber resembling a 'car wash', using air and electricity to 'clean' dust from the suits as the EVA crew re-enter the airlock. By changing the airflow and air pressure slightly in the habitation module, the residual dust could be kept inside the airlock. However, if such a system were to be incorporated on the lander, it would add to the complexity and mass of the payload, and would therefore have to be proven to be a necessity before installation.

*Suit ports* A more practical idea already under evaluation at NASA is a rear-entry suit backed into a receiver aperture in an airlock 'outer wall'. The operator method of entry resembles that of both the Russian Orlan and the US ZPS-Mark III suits, by which the crew-member climbs into the back of the suit, the hatch is closed and, when ready, the suit is detached from the aperture, allowing free movement across the surface. The process is reversed for the return to the habitation module. These 'suit ports' could be installed on the side of the lander, in a larger 'porch area' with a small access ladder down to the surface level. This would leave the dust outside on the exterior of the suit, but it would complicate access for servicing and maintenance on the outer surfaces.

## DEVELOPMENTS TOWARDS A MARS EVA SUIT

Although no Mars EVA suit exists there is a wealth of written data discussing aspects of design evolution, system testing, materials, biomedical issues, simulations and project developments. By the late 1990s, interest in manned Mars missions was renewed – partially as a result of cooperation in the development of the ISS (primarily in the US, Russia and ESA). In 1999 a new International Space Technology Center was chartered for the purpose of developing the concept of a manned flight to Mars. A twenty-one-member control committee from NASA, Russia and ESA reviewed possible scenarios for the flight to and from Mars, the biomedical issues of such a long flight, and the development of hardware, including the surface exploration suit and the testing of support elements and technologies.

### Zvezda evaluations

The Russian company Zvezda developed a concept and designed a prototype (or functional mock-up) without a life support system or outer protective garment. A semi-rigid rear-entry hatch design was selected, using elements of the Orlan suit to speed up manufacture and reduce costs. New hip joints were developed to assist in mobility, and facilities were developed for interfacing with an airlock or rover facilities. The knee and arm elements were improved Orlan soft joints, and the ankle joints and footwear were developed in cooperation with NASA and Hamilton Sundstrand in the US during 2000–01. These were tested and evaluated, and compared with earlier tests and evaluations. The tests included walking on horizontal and inclined surfaces, climbing ladders, kneeling on one knee, kneeling on both knees, bending forward, leaning back, using the pressurised glove, and retrieving items from the floor. Detailed photographs were taken of the movement ranges of elements of the suit, including the shoulders, arms, wrists, fingers, hips, legs, knees, ankles, feet and waist, both in isolation and in combination. Although the three male test-subjects highlighted some painful pressure points, improvements to the problem areas alleviated these concerns. Evaluation of the hard upper torso and helmet design showed that it was extremely convenient for donning and doffing, and had good visibility. There was a reasonable range of movement, and the wearer was more able to climb slopes and traverse flat and rugged surfaces. Tests were also conducted on ladders with steps up to 42 cm apart. Kneeling in the suit was comfortable for a considerable time, and the wearer was able to return to a standing position without assistance.[20]

### The I, D and H demonstration suits

In 1999, two American companies were evaluating the designs of what were classed as 'planetary suits', designed for surface exploration studies to formulate demonstration models for further test and evaluation. These suits were constructed by David Clark Company, of Worcester, Massachusetts (the D suit) and ILC, of Dover, Delaware (the I suit). The I suit was an improved version of the Shuttle EMU, which was much lighter and could be packed into a smaller volume. It also afforded better mobility and was less expensive. It featured a soft upper torso instead of the bulkier Hard Upper Torso (HUT) featured on the Shuttle EMU, which resulted in a suit weighing around 29.5 kg instead of the 45 kg of the Shuttle EMU system. As the idea was to evaluate planetary surface traverses, ILC worked with Redwing commercial boot-makers to adapt a commercially available boot for the demonstration suit, amending the sole and interior of the boot to include an air bladder inflation device to provide a snug fit. The ILC suit featured rotational bearings to add movement in the arms, torso and legs, allowing the wearer to rotate the shoulders and swing the arms and legs more easily.

By comparison, the DCC suit featured only one bearing (in the centre of the torso) and weighed just 12 kg. This allowed the upper body to rotate, and provided a means of access to the suit. Using the design of the Shuttle escape suit as a model, DCC redesigned the joints for more mobility and essentially produced a predominantly soft suit. The D suit was thus much lighter than the I suit, but it afforded less movement. The idea was to take the best out of what was learned from these prototypes – principally, the lightweight features of the D suit and the mobility of the I suit – and incorporate it into the next prototype. It was also a further step on the long road in developing an advanced EVA suit to replace the current Shuttle/ISS system.[21]

A third concept – the H suit – was developed by NASA Johnson Space Center in Houston, at the same time as the I and D suits. Using outside contractors, JSC engineers adopted a more rigid component, evolved from the early studies in the 1960s and 1980s, and resembling the I suit. The bearings on the H suit were in the same locations, but access into the suit was different. Tests of these suits were conducted at a simulated Mars landing site at Haughton Crater, on Devon Island, Canada, in 2000 – an ideal location for evaluation of a prototype suit in a crater-like terrain.[22]

Developing a Mars suit without a clear definition of how or when the mission would take place and exactly what hardware would be developed to support it is a challenge in itself, but it provides a long lead time prior to the definition of contracts to fabricate the flight models. Evaluating all concepts, materials, technologies, components and procedures helps eliminate impractical, overly technical and expensive proposals. Using current technology and applying it to future designs was a step-by-step approach that was used in the Apollo suit, and dozens of designs and models were produced before anything was developed for flight testing and operational use. The development of a Mars suit will probably involve exactly the same approach: determining what works, and making it cost effective, safe, reliable and easy to operate.

**Support technologies**

In addition to the suits themselves, work has also begun on the displays and control modules worn on the chest of the suits to inform the wearer of the status of the suit and its consumables. By the time the Mars missions are dispatched, a fully integrated computer (with heads-up display features) will include suit information, EVA plans, terrain maps, equipment or experiment information, 'to do' lists and communications and reference data, possibly linked to a satellite global positioning system for navigation and time-lining the actual EVA activities against planned ones. These devices could be as small as a wrist display but advances in such technology do not always yield a corresponding advance in the returns from the task. It might prove better to provide more simplistic data during EVA, allowing the astronaut to draw upon human experiences rather than totally robotic sources.

Other technologies under development include small sensors (coin size) designed to alert martian explorers to dangerous chemicals or potentially lethal radiation levels. Some of the experiments on robotic Mars spacecraft are designed to evaluate the environment near the surface and the chemical properties of the soil prior to sending astronauts there. Chemical sensors are large and bulky and normally intended for use in Earth based labs rather than for carrying around, especially while wearing a pressure suit. The new small sensors being developed by Nicholas L. Abbot of the University of Wisconsin and Rahul R. Shah of the 3M Corporation are designed to resemble small badges, without the need for electrical connections but which reveal visual indications of local toxicity.[23]

Studies by Hamilton Sundstrand Space Systems (HSSS) in support of the development of Mars suit technology have resulted in a conceptual design of pressure garment with a single wall, permitting heat transfer from the internal atmosphere to the external environment, a feature impossible in suits designed for a vacuum operation. This concept would use the presence of the thin Mars atmosphere as an aid to alleviating the body's natural metabolic waste heat. This means that a lighter or smaller back-pack could be used without the need for the specialised system needed to remove heat from the body. The suit was developed with the aid of information supplied by the 1997 Mars Pathfinder mission to Mars and other robotic missions to the Red Planet, and resembles the rear-entry Orlan type suit design. Segmented thermal overgarments could be put on or taken off the suit during an EVA to cope with varying surface temperatures of −193° F to +68° F. The suit would also be able to take advantage of the abundance of carbon dioxide in the martian atmosphere to extract oxygen, rather than transporting it to the planet and carrying it around in the back-pack, allowing for a lighter and more reliable support system. The suit parameters were classified as being able to support a 'basic' 4-hour unassisted EVA, but with two increments each of 4 hours possible, up to a maximum of 12 hours unassisted EVA capability as the traverses travelled further from the landing site or the complexity of the EVA work load increased.[24]

## MARS SUIT REQUIREMENTS

It is clear that an EVA performed on Mars will be very different from those previously conducted in space or on the Moon. Consequently the suits required to support human exploration of the Red Planet will have to be lightweight yet rugged, resistant to dust, easily stowed, comfortable over long periods of operation, fully reusable over many EVAs, adjustable for different sizes of wearer and capable of servicing and repair so far from Earth. In addition, any life support system will have to be lightweight, simple in design, use as many *in situ* resources as possible, allow for longer-term contingency usage and be safe and easy to repair or maintain. The exact design will also be affected by the choice of materials and their performance in the martian environment, the comfort of the wearer and, as always, the overall cost.

As well as the varied studies into the exact design of a suit and life support system for Mars, biomedical studies are being conducted to evaluate the human element of wearing and working in a pressure garment. These include the requirements for pressure, temperature, oxygen supply, rejection of heat and body wastes, refreshments, impact injury studies and prolonged physical activity both in shirtsleeve and suited conditions, and the effects on the bones and muscles after prolonged exposure to 'microgravity' or artificially induced gravity during the long flights from Earth to Mars and back again.[25]

A further study reviewed astronaut performances in wearing pressure garments and the implication for future spaceflight design.[26] This EVA research was being conducted by the MIT Man Vehicle Lab and was aimed specifically at the issue of suit mobility and future spacesuit design, utilising all previous experiences of suit design and operations by inputting the data into computers to provide computerised simulations of EVA, alleviating the need for time-consuming physical simulations and workplace limitations. Mathematical models were formed from the NASA Robotic Space Suit Tester (RSST) built by Sarcos Inc. in Salt Lake City, Utah and on loan to MIT. This is an anthropomorphic robot custom-built for NASA, featuring twelve hydraulically actuated joints on the right side and twelve posable joints on the left side.

The robot is known by the name M. Tallchief and can execute a preprogrammed series of moves, recording the position and torque of each articulated joint. The first phase of data collection was by unsuited human subjects who performed representative EVA tasks and their arm and leg joint positions were measured with an optical motion tracking system. This was repeated with suited test subjects and this data was then used to drive the unsuited robot in order to determine joint torques. Then, using a 4.3-psi pressurised EMU, the robot was programmed to repeat the movements. The second phase of data collection was a program of computer generated data that was then used in dynamic simulations of space suited astronauts and included a mathematical model, a physics model and dynamic simulations. With this improved understanding of astronaut movement, the physical limitations and range of the human form could then be applied to advanced space suit designs to assist in defining the fit and functions tests, materials selection, concept designs and subsystems to help determine the most effective suit design for future EVAs, including those at Mars, before a suit is fabricated and astronauts climb inside to test them. This drastically reduces the initial expensive development programme costs.

The idea that a Mars suit will be big and clumsy will not necessarily hold true. Varied developments are targeted towards reducing the bulk and mass of the suits and there will be studies into skin-tight coverings in future designs. Whatever the final design of the suit for Mars, it will be significantly different to those than have gone before. Dave Newman, one of the MIT researchers of the astronaut movement studies, observed: 'Everybody is used to spacesuits looking big and chunky, but we need to be more creative. I do not think we should constrain ourselves into thinking that a Mars suit has to look anything like the Apollo or Shuttle suit.'[27]

## REFERENCES

1 Gary L. Harris, *The Origins and Technologies of the Advanced EVA Space Suit*, American Astronomical Society, 2001, p. 422.

2 See Harris (ref. 1) for an expanded evaluation of what is required for fabricating an enclosure to support human life in a vacuum/partial pressure conditions.

3 Lieut Mike Thornton, USN, Dr Robert Randall and Kurt Albaugh, *Then and Now: Atmospheric Diving Suits*, *Underwater Magazine*, March–April 2001 (reprint).

4 Sam Barnes, 'Clothes that Clank', *Machine Design*, 17 August 1967, 28–32.

5 For further information on ADS programmes see ref. 4.

6 Development of Space Suits, reference file compiled by Lillian D. Kozloski, National Air and Space Museum, undated, post 1988.

7 See D. J. Shayler, *Walking in Space*, Springer–Praxis, 2004.

8 H. E. Ross, 'Lunar Spacesuit', *Journal of the British Interplanetary Society*, **9**, no. 1, January 1950, 23–37.

9 *Apollo Program Summary Report*, April 1975, Part 6, Flight Crew Summary; Section 6.1.2.7, Lunar Surface Operations, NASA JSC, Houston, JSC-09423.

10 C. C. Lutz, *Space Suits and Support Equipment*, NASA Manned Spacecraft Center Crew Equipment Branch, Crew Support Division, 7 February 1966; JSC History Archive, Space Suit Collection, University of Clear Lake, Houston, Texas.

11 Ref. 1, p. 127.

12 Ref. 1, pp. 128–162; also RX suit evaluation papers (RX-2 to RX-4) in JSC History Archive, Space Suit Collection, University of Clear Lake, Houston; Evaluation of RX-3 Hard Suit, HES-67-13, Human Engineering Section, MSC, Houston Texas, 3 May 1967.

13 Lillian D. Kozloski, 'Advanced Development Suits', in *US Space Gear: Outfitting the Astronaut*, pp. 145–168.

14 Ref. 1, p. 140.

15 Ref. 1, pp. 167–275 and 329–381.

16 I. P. Abramov and A. Ingemar Skoog, *Russian Spacesuits*, Springer–Praxis, 2003; D. J. Shayler, *Walking in Space*, Springer–Praxis, 2004.

17 I. P. Abramov and A. Ingemar Skoog, *Russian Spacesuits*, Springer–Praxis, 2003, pp. 277–285.

18 Ref. 1, p. 426.

19 Ref. 1, pp. 438–439.

20 I. P. Abramov, N. Moiseyev and A. Stoklitsky, 'Some Problems of Selection and Evaluation of the Martian Suit Enclosure Concept', paper presented at the 54th IAF Congress, Bremen, Germany, October 2003, IAC-03-IAA.13.3.02.

21 Greg Clark, 'NASA Plans Future Spacesuit for Planetary Missions', http://www.space.com, posted 10 November 1999,.

22  Harold Franzen, 'The Astronaut's New Clothes', Scientific American.com, posted 10 September 2001.
23  Robert Roy Britt, 'Safety on Mars', Space.com, 27 August 2001.
24  Tracey Guyer and Edward Hodgson, 'Hamilton Sundstrand Markets new ready-to-wear Mars suit', International Space Industry Report, 15 February 2000, p. 9.
25  Laura Parker, Cranfield University, UK, 'Design of an EVA Suit for use on the Martian Surface', presented at the 54th IAF Congress, Bremen, Germany, October 2003, IAC-03-IAA.13.3.03.
26  A. Frazer, B. Pitts, P. Schmidt, J. Hoffman, D. Newman, 'Astronaut Performance: Implications for future spacesuit design', a paper presented at the 53rd IAC, The World Space Congress, Houston, Texas, October 2002, IAC 02-G.5.04.
27  Scott Lafee, 'The Mars Collection', *New Scientist*, 2 September 2000, 34–38.

# Surface exploration

The primary purpose of sending humans to Mars is to conduct surface exploration. After establishing the infrastructure of the outward and return journeys, and designing the hardware (including EVA suits and support equipment), the surface traverses can be planned. As with other aspects of the first manned mission to Mars, there remains much to be decided, but some studies have investigated what might be expected over the course of the first excursions.

## POTENTIAL CREW ACTIVITIES ON MARS

Unlike the Apollo missions, the early Mars landings will probably feature simultaneous manned and unmanned operations – possibly including the control of robotic vehicles and automated investigations from either the manned lander or orbital vehicles. With the communication time-delays, direct Earth-based involvement in robotics will be much simplified – probably only supervisory or for program uploads and system reboots.

The exact timeline for surface operations will have to be determined when a flight programme is more clearly defined, but it will probably be considerably more than the few days of the Apollo missions. If the first landing is restricted due to safety considerations, however, then the timeline will be more compact (as with the Apollo missions). Depending on the mission profile and its timing, if the first crew was destined to remain at Mars for a month due to planetary alignment, then even the first landing would include a more ambitious surface programme than that of Apollo. No matter how many EVAs are planned, there will have to be the historically important and largely symbolic first human EVA on Mars before serious work begins.

If the case for a reasonable programme of EVAs is accepted, then what type of activity will be expected? Certainly with a longer surface stay, the potential for expanded activities will increase; but so will the hazards, difficulties and potential failures of equipment. Equally, a longer time on the surface at each landing site will be rewarded by far greater returns than the Apollo missions could ever have hoped for, despite the immense distance between Earth and Mars compared with the Earth–Moon system.

In 1988 an IAF Paper[1] examined what kind of activities the first Mars landing crew might be expected to complete.

- *Exploration* This could possibly include topographical mapping, reconnoitring for natural resources, and a general survey of the immediate area around the landing site.
- *Construction* Deployment of power sources for extended operations external to the landing vehicle would also require the clearing of access tracks through the rock-strewn local area. For an extended surface stay perhaps early habitation modules – previously landed, or constructed from the main landing vehicles – could be extended as a semi-permanent research base.
- *Manufacturing* One of the most probable scenarios of even the early landings will be the utilisation of the resources found at Mars to supplement, extend, or provide redundancy to delivered consumables or as a source for return-journey consumables. This would include water and oxygen production, production of other useful gases, metallurgy, and propellant production.
- *Food production* To supplement and eventually replace delivered foodstuffs, studies in agriculture, aquaculture and micro-chemostats will feature in these research fields
- *Maintenance* Ascent sequence servicing, habitat maintenance and prevention of system failure, housekeeping and equipment maintenance, repair, servicing and replacement.
- *Safety* Emergency drills, systems monitoring, repair and servicing, medical checks and procedures (exercise).
- *Training* Emergency procedures, new equipment and procedures, EVA simulations, ascent simulations, physical training.
- *Free time* Mars 'sports', personal investigations, 'philosophical reflection', contact with family and home, personal entertainment suite (audio/visual).

The paper also listed assumed scientific studies to be addressed in the first/early landings on Mars:

- *Geology* Studies in volcanism, including active volcanoes, seismic activity and Eolian activity. Studies of the evidence of water would also be included, exploring the channels, the permafrost and any water-laid sediment.
- *Atmosphere* Studies of the weather systems; photochemistry; climatology; analogies to any martian ice ages.
- *A search for life* Search for endolithic organisms; sulphur-based metabolism; evidence from beneath the superoxidised zone; various oases and warm, wet spots due to volcanic and impact processes; evidence of fossils (microfossils, unique structures and signs); survival of terrestrial organisms.
- *Relationship* Relationship, composition, and potential for resources; age and possible origin; the effects on the surface of Mars.

# LEARNING FROM APOLLO

Direct application of lessons learned from Apollo would be impossible at Mars, due to the completely different environments, the distance, and the time interval. However, there are indirect applications that could be applied to any future exploration of the Moon and, to some degree, at Mars. In 1992, Eric M. Jones – a laboratory fellow in Earth and Environmental Sciences at Los Alamos National Laboratory, New Mexico – and former Apollo 17 Lunar Module Pilot, geologist Jack Schmitt – a consultant at the Los Alamos laboratory – presented a paper that evaluated pressure suit requirements for walking on the Moon and Mars in light of experiences from Apollo.[2] The authors concentrated on productivity and the frequency of future EVA operations, based on the technology used.

During the Apollo era, the EMU and back-pack design operated for up to 24 hours without serious malfunction during exposure to the lunar environment. Within the context of Apollo, the pressure garments, related systems and equipment were very successful, resulting in a high productivity rate. On the other hand, experiences of minor equipment failures, crew fatigue, a few operational inefficiencies (mainly as a result of the particular design features of the suit used), and signs of wear and tear even during these short excursions, indicated that more evaluation would be required for extended operation both on the Moon and at Mars. The paper identified key areas in which further work would need to be addressed:

- Dexterity, mobility and fatigue of the user.
- Durability and weight of the EMU.
- Component susceptibility to dust abrasion.
- On-site cleaning, maintenance, and component replacement.
- Recharge fittings to extend the life of the hardware.

Operating inside the pressure garment in an alien environment for the first time will always induce an element of caution, but as experience increases, so too does the level of confidence in properly functioning components – and with this there will be a higher level of confidence to push the suits towards their design limits in order to achieve the assigned task. Such was the experience on Apollo. The first exit into the lunar environment (Apollo 11) came only four years after the very first EVAs in Earth orbit; indeed it was only the thirteenth EVA, on the ninth mission to attempt the exercise. Armstrong's 'small step' was also a giant leap of faith in EVA systems after such a short time. Now, as we look to a serious programme for exploring Mars, we can draw upon almost forty years of EVA experience and around 200 EVAs for planning a 20–30-year programme to prepare for stepping onto Mars.

During Apollo, the astronauts inflicted stress on each suit with repeated donning and doffing, crawling on their hands and knees in and out of the LM forward hatch, and moving on and off the LM. The back-pack's cloth covering showed signs of wear from rubbing on the seat structure, and the suit itself from using the lunar surface tools, deploying the experiments and handling large rock samples. The lunar dust also impregnated the gloves, the suit and the knees from handling, kneeling in and simply working on the lunar surface. Several of the lunar explorers, however, indicated that their confidence grew with each mission – particularly as their training programme developed. Indeed, on the Moon many of the later astronauts were more confident during their second and third excursions, based on their own experiences on the first excursion outside. They became far more appreciative of their situation during the occasional short breaks, when they realised where they were standing and what they were doing, and that their survival depended on a fully functioning suit. But throughout the seriousness of lunar exploration under the pressures of the Apollo timeline, they also took time to absorb the experience when possible, as they were unlikely to ever visit there again. Such a mixture of engineer, scientist, explorer, safety officer, and human being will undoubtedly feature among future explorers of Mars.

The overriding feature of EVA on the Moon was the ever-present effect of the lunar dust, which, in the low gravity, was easily kicked up, and settled on the suit surfaces and worked its way into joints and mechanisms. Dust abrasion wore away the outer fabric of the gloves, the slide mechanisms on the visors, the wrist and neck rings, the zippers and the connectors. Despite cleaning and lubrication, they still became clogged by dust, and were consequently stiff, and difficult to operate.

The problem of the dust was summarised by Gene Cernan, Commander of Apollo 17, during debriefing for the mission: 'Dust! – I think probably one of the most aggravating, restricting facets of lunar surface exploration is the dust and its adherence to everything, no matter what kind of material, whether it be skin, suit material, or metal, and its restrictive friction-like action on everything it gets on.'[3] He also said that by the middle of the third EVA (around 18 hours total surface EVA time), sample-bag locks and restraining locks on the LRV were failing to function, and that despite attempts to clean them they did not work properly. The effects of the dust on mirrors, cameras and check-lists was phenomenal. It affected averything on the lunar surface, and also inside the LM – but the crew had to live with it. It was on their hands as they cleaned off the suit, it was on their faces, and they were walking in it. No matter how careful they were, it found its way into every nook and cranny in the crew cabin and every pore in their skin, and Schmitt had some local respiratory problems. After they returned to orbit, the dust – now in microgravity – entered their eyes and throat as they cleaned themselves up for the transfer to the Command Module.

This has a direct application for any Marswalk, both in contamination and health issues, and will undoubtedly result in a controlled environment via a dustlock or airlock between the exterior suit lock and the interior habitation areas. The Apollo astronauts were on the Moon only for a few hours or days, but it is expected that on Mars there will be multiple EVAs over many months, and it will be very difficult to keep the inside of the vehicle clean throughout that time.

Wear on the gloves also came about from using the lunar drill. Cernan's right-hand glove was worn from using the surface drill to drill three holes, using a treadle to extract a deep core, and wielding a geological hammer at five geological stations. As he was right-handed, his left-hand glove was subject to less abrasion or wear except in the facing surfaces of the thumb and forefinger (indications of gripping).

Designing a dexterous hand assembly for a pressure garment has been one of the most difficult engineering tasks in spacesuit design.

During Apollo, it was not only the equipment that suffered as a result of the work levels and environment. The astronauts themselves also had to work against the suit design, resulting in a variety of minor hand and arm injuries. These included blisters and cuts on knuckles and wrists, fingernail damage, and soreness of the hands and forearms from having to counteract the inner pressure in order to squeeze their hands closed and grip the object that they were trying to grasp. There were various experiences due to carrying the ALSEP from the LM to its emplacement site. The two science packages, carried 'dumbbell fashion', were bulky if carried in the hands or in the crook of the arm. The astronauts' fatigued muscles recovered somewhat during 'overnight' rest periods in the LM, but they found that fatigue quickly returned on subsequent EVAs, to the point where their efficiency began to decline at a rapid rate – up to as much as 75%. It is interesting to consider what might have happened to the work rate on the planned extended Apollo missions, during which the crews would have been extremely busy during fourteen days on the surface. Apollo 17 astronaut Jack Schmitt said the experience of wearing pressure gloves while working on the Moon was like repeatedly squeezing a tennis ball in the hand for several hours!

In their paper, Schmitt and Jones offered one example of how working in a pressure suit forced the astronauts to work and move more slowly than a similar unpressurised activity exercise would have allowed. During Apollo 12, astronauts Conrad and Bean 'ran' about 100 metres towards the crater containing Surveyor 3. The 'running' consisted of a loping, foot-to-foot stride, allowing a top speed of about 6 km/h. The problem was that in order to achieve this, it took extra effort to deform the suit's waist in order to maintain the loping stride. As a result they quickly tired, and their heart rate increased to about 160 bpm. They therefore had to rest for a few minutes – which negated the advantage of their 'run'. By the time of the later Apollo missions (15–17), improvements in the waist area of the suit made them more flexible, so that the astronauts could comfortably sit on the LRV. As a result, the four astronauts on Apollo 16 and 17 were able to 'run' distances of more than 100 metres with no significant fatigue or increased heart-rates. Jack Schmitt adapted the Apollo 12 loping stride to produce a variation similar to that used in Nordic long-distance skiing. His 'run' of about 300 metres from the ALSEP site to the LM and across to another experiment emplacement area resulted in a peak heart-rate of only 130 bpm. That he was entirely comfortable with this was evidenced by his singing about 'strolling on the Moon one day ...'

Due to the new design of the suits it was easier to kneel than it had been for the Apollo 11–14 astronauts, although it was still not a simple matter of kneeling down and standing up again. Because of internal pressure, the undulating surface, the bulk of the suit and its back-pack, and restricted vision, all of their movements had to be carefully thought out and conducted with care. Loss of balance was an ever-present possibility, and several astronauts had their antics preserved on film. In viewing these films it becomes evident that apart from the apparent danger of an injury or a torn suit, the disturbed dust coated instruments, experiments and camera lenses, and was a constant problem. If an astronaut fell over, it was both time-consuming and tiring to stand up again – whether by using a 'press-up' or a long-handled tool, or by being helped by other crew-members.

A final point made by Schmitt and Jones related to the useful lifetime of Apollo EVA hardware. The Apollo missions were all of short duration, with no more than three surface EVAs of 8 hours each on three consecutive days. The suits were tailor-made for the individual astronauts, and were worn during ascent, rendezvous and docking, lunar descent, EVA, lunar ascent, and re-entry. Their operational requirement was therefore far more than their intended use for EVA. No more than three 8-hour EVAs were planned on the last three missions, and no operational data are available to indicate how long an Apollo lunar surface EVA suit could have remained functional. But the consensus among those who used them is that, without a major failure, they could not have survived more than twelve 8-hour excursions before an unacceptable level of leaks or accumulated minor failures rendered them dangerous for the occupant. In addition, Apollo astronaut debriefings indicated that in order to reach this ceiling, additional inter-EVA time would have to be allocated to maintain the suits with thorough cleaning, seal lubrication and minor repairs, as well as allowing the crews to rest.

Such maintenance would also have required working areas away from the crew habitability area – which was impractical for Apollo, as the LM was too small and the EVA programme too crammed. Before the Apollo landings it was impractical to do much about suit maintenance, but the later missions at least afforded the crews time to doff the suits and use the ascent engine cover as a work area to clean the suits by wiping the enclosures clean and adding a coating of lubricant. Care had to be taken to avoid transferring debris from one suit to another in the confines of the LM cabin.

Some of these lessons have filtered down to the later programmes, for which dedicated EVA preparation and storage areas have been designed, and an EVA day–rest day–EVA day cycle incorporated into the flight plan on later Shuttle missions. During the Skylab missions, the increased volume of the OWS allowed the suits to be dried, vacuumed and cleaned, and the Russians have accrued more than fifteen years' experience of on-orbit suit management and maintenance. Combined with many years of experience in Shuttle and ISS suit management and care, this will be useful for Mars missions. However, the continuing overriding factor about Mars (and the Moon) is dust – and on Mars there is the additional problem of corrosion by the atmosphere.

It is clear that an Apollo-type EMU design for use in deep space or on Mars would be a limiting factor in the duration and number of EVAs that could be completed with each suit system. In their paper, Jones and Schmitt assumed a maximum life of ten EVAs for each 80-kg Apollo EMU, which indicated a degradation of 8 kg/day. The bulky EMU design would therefore not be practical on extended missions lasting many months. Due to the limited internal volume of a spacecraft, it would probably be impractical to carry more than three complete EMUs per person (prime, back-up and reserve).

Using the Apollo experience, the authors then considered (from 1992 perspectives) a possible permanent lunar base and expeditions to Mars, noting that a surface stay of several months on the planet was plausible, and that EVA operations on Mars would feature both exploration and the construction of facilities, probably requiring daily surface activities in the first few weeks – a daunting prospect after a long flight from Earth. Appropriate back-up facilities and spares would ensure that EVA operations were unhindered by malfunctioning equipment, but this would also increase the payload mass and restrict the volume inside the lander. Their suggestion that the design life of most components of planetary suits should exceed the number of planned EVA hours was sensible, and was set at two hundred 8-hour EVAs for heavy components. Advances in technology, variable wear on components, and partial replacement of elements, would also help in extending the life of the suits. The authors exemplified the Apollo 17 mission, during which fingerless palm cover gloves were worn during EVAs 1 and 2 and then discarded for EVA 3. These protected the palms from abrasion, leaving them relatively unscathed for EVA 3. Perhaps at Mars, a supply of small, lightweight over gloves, replacement outer visors, knee pads, replacement fans and pumps (if included in the final design of a Mars suit) would be included in a baseline EVA suit maintenance kit.

Apollo clearly demonstrated that increased mobility in the suit joints and dexterity in the gloves greatly increased the productivity of the astronauts and reduced the possibility of fatigue and injury, even though it was still hard work to 'fight' against internal pressure. One suggestion put forward was to design suit gloves and upper limb areas with multiple 'rest' positions, rather than fully relaxed or full strength. Between each EVA, the astronauts would resupply and recharge systems in the back-pack, with longer life and cartridge-type replacement helping to shorten time-consuming housekeeping work and giving the astronauts more time for planning and rest.

Safety is another element that will play a major part during Mars EVAs. Apollo had the walk-back rule in the event of LRV failure, and the capability to share cooling water should a pump fail. Fortunately, this remained an unused option, but it left open the question of the effectiveness of such a system on the lunar surface and in keeping an astronaut alive.

With plans for initial excursions around the lander on Mars, a similar system will probably be adopted; but even on the first landing, a stay of several weeks or months might include some longer excursions several days from the lander, providing that safety systems and hardware support was in place. In this case, remote field resupply, recharging and contingency supplies will add to the burden of rover and equipment design.

It was clear that the Apollo technologies, while useful, would not be practical for the Mars programme (or, indeed, a return to the Moon for an expanded programme). At Mars, the mass of the EMU will need to be significantly reduced, hard-shell technology would need to be incorporated to increase durability, reliability and maintenance, and the 3.8 psi pressure that, on Apollo, hindered limb and hand movements over many hours, would need to be reduced. The key to successful Mars EVAs will be the development of an operational and extended-duration surface suit system for repeated EVAs, with flexibility, mobility, and minimal servicing requirements. The other key element is surface planning which is both achievable and productive, while at the same time retaining flexibility in real-time situations and ensuring a high degree of crew safety.

## A STUDY IN SURFACE ACTIVITIES

One of the studies of surface activities for an early Mars landing crew was commissioned by NASA from a combined team of former Apollo and Skylab astronauts and a range of specialists from the fields of human factors and biomedical aspects of extended spaceflight.[4] The team suggested that a surface stay of 90 days would be better than a 20-day or 40-day stay, given the substantial commitment to send humans to Mars. Using contemporary NASA guidelines, the team suggested a mixed human and robotic extended-duration exploration programme rather than a brief stay before the flight home – a true scientific expedition rather than a short visit. This essentially combined all the achievements of Apollo from six landings and what was planned for the Moon in the 1970s, adapted to the exploration of Mars during one mission. In this detailed 1989 report, the mission was targeted for a 2003 window, and would be flown on a minimum-energy direct trajectory (conjunction), with a single piloted vehicle carrying both crew and cargo over a period of 32 months. A total of eighteen months would be spent in Mars orbit, with at least 90 days (and perhaps more) on the surface. This may seem a short duration on the surface for such a long stay at Mars, but prior to landing it would have included an initial planning stage in orbit (to determine surface conditions at the landing site, similar to the Viking unmanned missions). This would provide the landing crews with sufficient time to plan their EVAs in detail, as well as time for the changeover period and contingency planning in the event of planet-encompassing dust-storms. In this study, a crew of eight would be divided into two teams of four, alternately descending to the surface to achieve two landings on one mission.

From orbit, the crew would conduct precursor mapping, and atmospheric and surface observations and analysis. A group of robotic explorers would be operated via telepresence, assisting with the final landing site selection and detailed mission planning, and a network of orbital communications satellites would be deployed for constant orbital communications relay from the surface to the orbital craft, even when over the local horizon. The orbital crew would also undertake support and instruction roles for the surface crew (essentially a mission control) in lieu of direct communication with Earth.

Their EVA tasks would include a variable programme of durations and ranges totalling 600 man-hours per site (1,200 man-hours for the mission). The surface crews would explore the terrain, collect samples, deploy geophysical and atmospheric monitoring networks, complete geoscience investigations and an extensive photodocumentation programme, conduct a range of target-of-opportunity and follow-up investigations, and set up communication stations.

The study evaluated trajectory operations and operational considerations, which would affect the duration of the flight to and from the planet, the length of time available for staying there, and the medical condition of the crew on arrival at Mars. Operational issues for EVA addressed by this study reflected the extreme differences from Apollo operations, in which hardware was flown with only days or weeks between system-testing and missions. For Apollo, the overall mission exposure of the equipment to the extremes of the environment (thermal conditions, radiation levels, vacuum, g forces) was brief, and any long-term effects were negated by the short duration of each mission (no more than twelve days from Earth). In contrast, Mars mission equipment would have to survive for hundreds of days before it reached Mars and for hundreds of days on the return journey, plus the time spent at Mars. EVA equipment would not wait for only three days before operational use, but for at least 6–9 months, and there would therefore be an increased risk of equipment malfunction and a greater need for maintenance schedules. Due to the complexity of the equipment, the crew might not be able to repair or maintain some of it *en route* to Mars. Unlike on the Moon, there would also be the added influences of surface environments (wind, dust-storms and corrosion), which would also affect the landing and ascent profiles and the landing craft. In addition, instead of the three days on the Moon, the equipment would have to maintain life support and surface equipment for weeks, or possibly, in the case of an emergency, for several months. In this case, a variety of contingencies might arise during surface operations to significantly change the planned exploration programme:

- Delays in descent caused by vehicle malfunction, severe atmospheric or weather conditions, surface conditions, problems with the main spacecraft, or the overall health of the crew.
- Delays in surface exploration activities resulting from problems with EVA hardware or support equipment, unfavourable weather, crew adaptation from microgravity to Mars gravity, or difficulties in preparing to support EVA operations.
- Delays in the ascent from the surface due to vehicle malfunctions or adverse weather conditions.
- There would also need to be a contingency programme to allow premature ascent at any reasonable time, due to degrading systems and contingencies with the crew.

The selected trajectory and equipment chosen at the time of the mission would have major effects on these issues.

**Crew-member suggestions**
In the 1989 report it was suggested that surface exploration could be achieved by two teams of four astronauts staying for 45 days each and completing about 300 man-hours of EVAs, varying in duration from 2 to 10 hours. As one team of four descended and explored the surface, the second would remain in orbit to act as 'mission control', exchanging roles for the second landing, with the additional advantage of the first landing crew, now experienced on the surface, acting as mission controllers on orbit during the second and probably more advanced landing and exploration programme. The eight crew-members would ideally include:

- Two commander–lander specialists (with back-up habitat specialist skills).
- Two co-pilot–habitat specialists (with back-up commander skills).
- Two physician–EVA specialists (with back-up duties as geologists).
- Two geologist–EVA specialists (with back-up physician skills).

At least two of the crew would be qualified in astrophysics, and would carry out long-baseline observations to and from the planet. There would probably be one overall mission commander for the eight crew-members, with the second being his back-up, and each landing crew of four would assume prime responsibility for one of each of the four categories, offering the best advantage of all disciplines and back-up skills in orbit.

In comparative studies at the time, the Office of Exploration suggested an eight-person crew, with four specialist landing crew-members and four specialist orbiter crew-members, while JSC studies indicated a three-person crew, with two descending to the surface for a short period of EVA, and one remaining in orbit. The advantage of the 1989 study is that it utilises all eight crew-members, with back-up, offering all eight the chance to experience the additional excitement and achievement of walking on Mars and to perform useful support roles in orbit after such a long flight.

### Initial surface activities

Advocating a cautionary approach, the study proposed an investigation and adjustment phase that would precede any human surface activities for perhaps up to seven days on the surface. While the crew remained in the lander to fully adapt from microgravity to Mars gravity, a complete systems test and check-out of equipment would be completed and on-site environmental studies conducted, with the flexibility to deal with on-site environmental conditions not foreseen on orbit prior to landing.

Unlike with Apollo, an immediate EVA shortly after landing would not be advisable after such a long exposure to microgravity flight conditions. Any initial EVAs would also be restricted to the immediate vicinity of the habitat/lander for safety reasons, reducing the chance of any nasty surprises on the surface. In extreme cases of hazards or difficulty detected prior to embarking on EVA, a more constrained contingency EVA, plus post-EVA operations and sample-handling activities, could be mounted inside the habitat, offering flexibility for surface operations and reducing the need for a large mass of geological samples to be returned to Earth.

### Mission control in orbit

An eighteen-month 'residency' in orbit around Mars would provide ample time for final landing-site selection and detailed planning for surface operations, depending on the state of the weather at the time. By placing two landing crews at different places in the same region during one mission, a certain range overlap might be possible to fully explore a given region of the surface for a more detailed analysis of local topography, environmental issues and geology than would be possible on one site or two far-removed landing sites. The use of robotic exploration devices between sites would be an option, allowing extended use of the same hardware and recovery of samples from the 'in-between' sites. A potential risk in using the second back-up lander would be the need for an alternative contingency to 'rescue the second crew' in the event of a systems failure on the second lander. This might be a contingency option either incorporated into the second lander, or a separate purpose-built vehicle without the capability of extended surface-stay time, but carried as an additional second level of redundancy in the event of main lander failure.

This two-landings-per-mission approach would provide a quicker, more extensive and deeper understanding of the requirements at Mars before the establishment of more permanent habitats.

**Why 90 days?**

The 1989 study team determined that 90 days would clearly be a more suitable surface stay-time than the 20–30-day stays which were being suggested at that time. The final report stated: 'Put simply, if we go to the trouble and expense of sending highly capable people to Mars, we want to maximise the probability of success and get the most from the effort ... We want to do to much more than raise the flag, have a quick look around, take a few pictures and leave tracks in the martian soil. The human exploration of Mars is not an exotic field-trip like a trek to the North Pole. Rather, it is the first scouting of another planet, a phase that may lead to eventual settlement on Mars and movement further into the Solar System. Seen in this evolutionary perspective, the initial human mission to Mars should be a concentrated, detailed survey in the tradition of the Lewis and Clark expedition or Darwin's survey of the South Pacific. We cannot do that kind of detailed survey in sixty EVA man-hours over twenty days.'

Ninety days on the surface would allow exploration of two sites for 45 days and around 300 man-hours of EVA at each. With such a duration, the lander crew would also be afforded sufficient time to deploy monitoring equipment and verify the proper functions for long-term study after they return, and they could perform necessary repairs and make adjustments to further maximise the operational programme and improve the quality of data while still there and after they leave for Earth. Forty-five days would also allow the crew to 'fully examine and sample complex scientific objectives, to follow up on observations and unexpected results, to go back to an interesting geological formation for a second or third look, to confer with experts back home, and to do interactive science in this new environment.' A brief, 'quick-look' mission would not permit these opportunities.

Sending two alternate crews to two different sites would also double surface experience on one mission, and would provide an alternative source of data to just a single landing area. The problems would be in providing duplicate sets of hardware and consumables to support such a profile, or ensuring that the equipment could last for the entire mission. Psychologically, the advantages of spending 45 days in a gravity environment between the long flights there and back are obvious, and proving that it could be achieved twice would be an immense boost for the morale of the crew, especially as they would all be offered the chance to walk on Mars.

**SURFACE EXPLORATION: 1989 STUDY PLANS**

By allowing each crew to remain on the surface for 45 days they would be able to adapt to partial gravity midway through the mission, and it would increase the productivity of all crew-members, and not just the landing crew. The study recognised that, unlike with Apollo, the crew would not simply put on their EVA suits and step outside a few hours after landing and months after leaving Earth. Estimates indicated that acclimatisation from microgravity to Mars gravity to reach peak activity may take as long as two weeks, but using data from Soviet and Russian long-duration residency crews on Salyut and Mir, the study team recognised that countermeasure activity *en route* might help accelerate readaptation. Clearly, however, an immediate 6-hour EVA programme would not be advisable or practical.

Of their first three weeks on the surface, the first week would consist of adaptation to the new gravity environment and preparations for future surface activities conducted from inside the habitat. It would probably also include the deployment of a back-up or contingency automated science station and a robotic surface vehicle to obtain back-up or contingency samples. During week 2, human surface activities local to the lander would gradually increase in alternating EVA teams of two. In week 3 they would establish the first extended EVAs away from the base – again by alternating pairs – and weeks 4–6 would build upon the extended-duration surface exploration phases begun in week 3.

As previously mentioned, several other reports proposed a 20-day surface stay. This would be the equivalent of the Apollo approach in which almost every minute of EVA time would be important. Readaptation, however, would cost some EVA time in the early days as the crew regained fitness and stamina, so a 20-day sixty 60 man-EVA hour surface stay would not equate to much more EVA surface exploration than the whole Apollo programme achieved during six landings on the Moon.

**Definition of a Mars EVA task**
An overall definition of any EVA is a combination of operational requirements, the capabilities of the human crew, and the constraints of the environment in which they are working. In the 1989 study, the team listed the difficulties at Mars as 'a hard vacuum; ubiquitous, permeating and potentially abrasive dust; extreme seasonal and daily temperature variations; high winds; dust-storms; solar radiation storms; the presence of local low-traction and low-bearing-strength terrain due to dust accumulation; and a slightly oxidising surface material environment.'

In addition to these 'environmental' constraints, there is the need to provide adequate environmental control and life support capability for several hours, with contingency and back-up facilities, effective communication and information technology support, an effective mobility and reliability capability with the suits, and a range of EVA support hardware and facilities that together can offer a reasonable chance of mission success, balanced with a productive workload and a very high degree of crew safety.

For operational constraints in the planning for any EVA, a number of questions need to be addressed relating to the task (or tasks) to be performed:

- What are the objectives of the task?
- What is required to achieve these objectives?
- What can the EVA crew do to achieve this task?
- Can they achieve this within the constraints of the hardware, timeline and safety?
- If not, why not? – and how can they be assisted by using associated means such as remote or robotic equipment?
- How much time should each task occupy, and can it be achieved in one EVA or multiple EVAs?
- What effect will the success or failure of the EVA task have on other elements of the surface programme? And are there contingencies and back-up procedures in place in case of primary failure?

**Planning the EVAs**

The exact timeline of the EVAs will depend on the hardware and procedures developed for the real mission to support the surface activities and achieve the objectives decided upon within that programme. However, regardless of these definitions there remains a generic list of EVA tasks that are applicable to any surface exploration of the planet,[5] including:

- Activation and deactivation.
- Adjustment.
- Assembly and disassembly.
- Calibration.
- Check-out and confirmation.
- Cleaning and decontamination.
- Collection.
- Communication.
- Connection and disconnection of electrical, mechanical, fluid, gas and data links.
- Definition.
- Display.
- Documentation by verbal, photographic and video means.
- Equipment, experiment and hardware servicing, maintenance, repair and reconfiguration.
- Labelling and encoding.
- Loading and unloading.
- Maintenance tasks and performance.
- Monitoring.
- Navigation.
- Observation and detection.
- Positioning and alignment.
- Setting up.
- Stowage and unpacking.
- Testing.

- Transfers.
- Traverses.
- Verification.
- Visual inspections.

Extensive EVA training and simulation will have been accomplished by the engineers, scientists and astronauts prior to embarking on the mission, and this has already begun in the pioneering work completed by the Mars Society and NASA-supported simulations. In addition, it is inevitable that flight-testing of elements of hardware will take place in space (whether this is at the ISS or on the Moon remains to be seen), and refresher training will take place during the flight to Mars, but surface exploration simulations (on Earth or on the Moon) will undoubtedly be carried out long before the months of flight from Earth to Mars. The experience of the Russians on long-duration space station missions showed that it is difficult to perform a task that was trained for weeks or months earlier on Earth, but not tried out during the mission (the 1997 Progress collision with Mir is a case in point), and keeping a crew trained at peak performance in light of Shuttle delays has been a problem for crews and trainers over the years. A Mars mission will be no different. The time between performing a suited full EVA timeline simulation on Earth or on the Moon and carrying out real EVA on Mars will result in new challenges and problems.

In their evaluation, the 1989 study team proposed a methodical exploration programme over the 45 days that offered success with mission objectives (defined from orbit in real-time prior to landing) and allowed the crew to fully adapt and gradually increase the range of their EVA operations while still retaining a high level of safety. The suggested 45-day surface stay offered a total EVA plan involving a 360-degree range divided into four traverse sectors, addressed from the third week on the surface onwards:

*Week 1* A programme of post-landing checks and reconfiguration of the lander or habitat and a period of readaptation by the crew, involving medical checks, physical exercise, controlled diet and planning sessions. A series of hardware preparations and checks and the initiation of remote and automated data-gathering and sample selection local to the lander, as well as back-up experiment deployment. Depending on the chosen final mission profile there might be a few short preliminary (and ceremonial) EVAs to configure the lander for an extended stay or to reach a separate habitat if that is the configuration.

*Week 2* In the opinion of the 1989 study group, this was the most effective time-period for the crew to begin surface activities, remaining locally around the lander (similar to Apollo 11–14), where a more extensive surface experiment package could be established, and where detailed examination of local surface features could be conducted and samples gathered, along with planning for more extensive EVA forays the following week.

*Week 3* Remote rovers could be deployed on the first few days on Mars, but the study group suggested that during this week of operations the range of the rovers – both robotic and by human – should be extended for exploration further from the

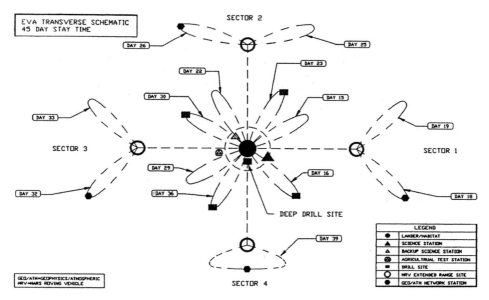

EVA TRANSVERSE SCHEMATIC
45 DAY STAY TIME

SECTOR 2

SECTOR 3

SECTOR 1

DEEP DRILL SITE

SECTOR 4

GEO/ATM=GEOPHYSICS/ATMOSPHERIC
HRV=MARS ROVING VEHICLE

| LEGEND | |
|---|---|
| ● | LANDER/HABITAT |
| ▲ | SCIENCE STATION |
| △ | BACKUP SCIENCE STATION |
| ⊕ | AGRICULTURAL TEST STATION |
| ■ | DRILL SITE |
| ○ | HRV EXTENDED RANGE SITE |
| ● | GEO/ATM NETWORK STATION |

Schematic from the 1989 study report showing the systematic exploration of the surface of Mars over several weeks.

lander but still within walking distance (as during Apollo 15–17). The landing area would probably be divided into traverse sections of medium- and long-distance targets to be achieved in four parts (probably north, south, east and west of the lander or landing site).

*Weeks 4–6* This would be the period of continuous extensive surface activities from the lander, within hardware and operational constraints (similar to what was planned in the 1960s with pressurised Mobile Laboratories (MOLABS) on the cancelled 'Apollo' follow-up missions in the late 1970s). There would also be a revisiting and contingency EVA capability to close out or extend specific EVA findings in real time.

**Achieving the objectives**
The ideal landing area would be favourable for multiple exploration objectives, and would be as close as possible to the primary objective area. Initial surface exploration would be carried out close to the lander. The crew would then gradually extend the range of operations as they became more experienced and more adapted. The radius of these exploration sectors would be determined by the mission profile, the geographical features of the landing area, surface and climate conditions, the limitations of the life support system, the constraints of the mobility systems, and the performance of the crew and equipment.

*EVA duration* In adopting the 45-day surface stay, the team recognised the need to maximise productivity and EVA time, while at the same time incorporating adequate rest, maintenance and planning time (which was not incorporated on the short Apollo forays on the Moon). Typically, an EVA would last about 6–8 hours, but with technology capable of supporting 10 hours in emergency or contingency situations. Initial EVAs might only last about 2–4 hours as the crew become accustomed both to the surface conditions and to operating their equipment on Mars. Experience with Shuttle EVAs showed that performance is maximised by rotating EVA crews and providing adequate rest days in between, although the nature of Shuttle missions is such that the EVA rest day is usually spent preparing equipment or supporting the other crew on their own EVA. The 1989 study suggested an alternative EVA crewing system to overcome this, with two non-EVA days before and three non-EVA days after 10-hour EVA periods.

*EVA real-time planning* Appropriate time would need to be incorporated into the flight plan for the analysis of results, future activity planning, and for frequent ascent simulations as part of the safety and contingency training programme. The first phase of surface activity would be based on simulation and robotic programme findings (especially the examination of automated sample return missions to determine unforeseen hazards in the environment), and would include adequate time for adaptation to Mars gravity, the deployment of scientific instruments, the gathering of samples (atmospheric and geological), observations and surveys, photographic and audio–visual data-recording, and taking advantage of planned and unplanned targets of opportunity, as well as time to revisit interesting sites more than once or twice, and contingency provision for any unforeseen circumstances.

*Second landing EVA refinements* The benefit of incorporating two landings on one mission would be that the experiences of the first crew returning to the orbital craft would be used to update and brief the second landing crew. During this interim, the combined crew would probably conduct a period of sample analysis, a series of debriefings, and servicing of hardware. Then, while the second landing crew incorporates any modifications prompted by the experiences of the first crew, remote operation of the rovers would continue the scientific return on the surface until the arrival of the second crew. The landing and surface exploration profile of the second landing would be similar to the first, but would be significantly enriched by their recent experience, and adjusted accordingly. At the end of their 45 days on the surface, the second crew would return for their own period of orbital sample analysis and debriefing, terminating remote telescience data collection or switching to sending telemetry to Earth from the surface experiments and surface vehicles still in operation.

The return journey to Earth might include visits to Phobos or Deimos (with surface exploration by automated or manned means), a fly-by of Venus, and continued scientific studies of the retrieved samples and astrophysical objectives, as well as medical tests and physical exercises similar to those carried out during the outbound journey.

*Rovers* Automated rovers proved very successful on the Moon, and at Mars they will undoubtedly supplement the manned phase of surface exploration. Apollo also demonstrated the benefits of using a roving vehicle compared with a foot-traverse programme, and the local survey vehicle will be an integrated element on Mars – probably on the very first landing. For more extended surveys, a pressurised module would perhaps be required, but on the very first landing these might not be used, as it would require a guaranteed element of contingency and rescue capability. Until the experience level has increased on Mars, expeditions venturing days or weeks from the landing site or base would not be advisable. The designs of such rovers and scientific survey vehicles await the future programme, based on data received from current and forthcoming automated exploration programmes.

If, as the 1989 study suggested, a pair of Mars Roving Vehicles are utilised on the first landing, with remote as well as manual driving facilities, an unmanned rover could be sent out to a more distant objective as the four sectors are explored, and parked halfway, using the rest period to recharge its batteries. This could take place on the day before the crew take the second rover to join it and then transfer to the first rover to continue their EVA to the objective, leaving the second rover to recharge its batteries for the trip back to the lander. This so-called 'pony express' option offers extended EVAs, but with each rover venturing no further than walk-back distance from the primary objective to the parked rover and from the parked rover to the lander. More advanced expeditions of several days would probably be carried out on the third or fourth Mars missions, and here a pressurised rover would offer life support capabilities and rescue contingencies for a journey of several days, independent of the lander or base.

*EVA Robotic Assistants (ERA)* In addition to manned roving vehicles, studies have already been initiated into providing an EVA robotic assistance vehicle for planetary surface exploration.[6] In September 2002, field tests were completed on an Earth-based prototype designed to test human–robot interaction and software architecture. Such a robotic assistant is able to:

- Transport tools, equipment, geological samples, and so on.
- Track and follow astronauts during traverses.
- Provide a remote roving base station for additional safety, and supply information about the traverse via imagery and associated data.
- Respond to voice commands from the astronauts.
- Respond to direct commands from the lander or base.
- Assist the astronaut team by deployment of science instruments, or provide an autonomous roving explorer independent of astronauts.
- Secure advance scouting information and support investigation of difficult terrain.

Regular tests have been carried out on such vehicles at the simulated planetary surface area located at NASA JSC in Houston, Texas. During additional field tests (near Joseph City and Meteor Crater in Arizona), 'astronaut' test subjects – either shirtsleeved or wearing pressure garments – performed simulated traverses which included the manual deployment of a geophone to record seismic activity. The ERA supported these simulations by pulling a science trailer, following the astronauts, and deploying the geophone automatically – all of which demonstrated its independence. Using a laser rangefinder, the location of each human was established and a requested distance maintained as required. This type of tracking of the astronaut team was also performed during night simulations. It was suggested that in future trials, each test subject should wear an 'infopak' containing a GPS unit and a range of biomedical sensors, allowing the ERA to monitor the medical condition as well as accurately tracking the position of each astronaut to allow exact plotting of their movements and station-stops during traverses – all of which would improve their safety.

The ERA test-bed is a four-wheeled, skid-steered rigid suspension base, although a more advanced four-wheeled independent drive base with suspension has been built and has begun to undergo tests. The upper deck is modular in design, and houses sensors and processors mounted on the mobility base. Initial tests utilised battery-powered systems, but experimental work is being conducted to support a fuel-cell option for extended operations. A remote manipulator system attached to the ERA can reach storage locations on the unit, and it can also be used for obtaining samples or the deployment of instruments and the retrieval of items. The arm – designed by Mertrica Inc – incorporates multiple attachment points, and its end effector is a three-fingered 'hand' developed by Barrett Technology Inc.

**Operational requirements**

Several parameters need to be addressed prior to attempting human exploration of Mars. The method, timing, procedures and hardware required to support such a long mission would need to be ascertained, but the 1989 study team also discussed the following:

- How far could the astronauts venture from the lander, and what mobility, navigation and communication systems would they require?
- What radiation protection measures are needed, and what limits does this place on repeated surface EVAs and total surface exposure?
- What is the optimum EVA duration, including pre- and post-preparation time, depending on the chosen final design of surface suit hardware?
- What are the work parameters: how long can the crew work, how far can they walk in accordance with safety and physical conditions, and how much robotic and automated equipment is required to support these activities? Identified areas of further discussion and development were: the design and capabilities of the EMU and life support unit; EVA tasks and sub-tasks; the gravitational readaptation of the crew; the physical and physiological condition of each crew-member as an individual and as a team-member; the geography and surface features of the landing site; ionising and non-ionising radiation exposure levels; day–night cycles; dust-storms; temperature and thermal extremes; equipment corrosion and 'weathering'/degradation; the number of EVA crew-members required for a remote and difficult task (such as removing stuck drill cores, repairing major equipment, and so on);

the degree of automation and teleoperation available to support and supplement human EVA operations; the design of rovers or mobility vehicles (Mars flyers?); and the status of the orbital vehicle.

- How long is the Mars working 'day'? How long can even a paced EVA programme be sustained and still maintain a high rate of productivity?
- Mars duty cycles.
- Optimising the duration of EVAs.
- Mobility and transition on Mars.
- EVA rescue capability.
- Shelters.
- Dust locks.
- Dust-storm contingency.

**Hypothetical scenario**

The following EVA plan, based on the findings of the study, was published in 1989:

*Week 1 operations (surface days 1–7)* Following a successful landing, the crew will proceed through a programme of 'stay' and 'no stay' decisions based on their own medical condition, the condition of the landing vehicle, the stability of the surface under the lander, and local weather. Apart from descriptions of the views out of the windows, they will prepare for an immediate take-off if an abort or contingency situation suddenly arises. Upon the decision to remain on the surface, which will probably be updated several times during the first hours, the crew will reconfigure the vehicle from descent mode to surface residency mode and begin their programme of adaptation to gravity. About an hour of medical exercise and monitoring will be carried out on the first day, and, depending on the final mission timeline, robotic and automated devices may be deployed to obtain short-stay data and samples in case of a sudden need to depart. Over the next six days, the medical exercise will be increased by about 15 minutes each day up to 1.5 hours by the sixth day, as the crew prepares the equipment for EVA and reconfigures the inside of their lander/habitat for 'residential' mode. The lander will also be prepared for extended laboratory studies to support surface operations and examination of samples. A programme of communications sessions with the lander will be established for mission support conferences with the specialists on the orbital craft and Earth, for public relations exercise, and for private communication with families. Day 7 will essentially be a day of rest, apart from housekeeping, an extensive medical examination, and probably a review of remote surface activities and refinements to the planned EVAs over the following week.

*Week 2 operations (surface days 8–14)* After about two hours of preparation with all four landing crew participating, the first two EVA-designated crew-members, wearing full pressure garments, will exit the lander on Day 8. Depending on the final design of the suit and the procedure for leaving the lander, the first person will descend to the surface to take their own 'giant leap' into history. By the time the first person walks on Mars, suit design will probably allow a total EVA duration of about 10–12 hours, with 6–8 hours as an operational duration; but on this first foray on the surface, the EVA will probably last only 2–4 hours. This will be sufficient to deploy large communications antennae and the early elements of the Mars Surface Experiment Package (MSEP), and for the first collection of surface samples by humans. The intravehicular crew-members will act as EVA CapCom, choreographers and photodocumenters, and will organise the relay of TV pictures back to the orbital craft and Earth. There will also undoubtedly be a programme of ceremonies to talk to world leaders, deploy flags and plaques, and remember fallen colleagues. After completion of their surface tasks and any 'get-ahead' objectives, they will return to the habitat and complete a programme of post-EVA activities by recharging the life support system and cleaning the exterior of their suits and other equipment taken into the airlock, followed by a debriefing with the orbital crew and probably with support scientists and engineers on Earth. After taking off their suits they will re-enter the habitable section to celebrate their success. This will take up to 3–4 hours, before a well-earned rest. With Marswalk One completed, the age-old desire to step onto another planet will have been achieved; but the exploration of that world will have only just begun.

*Day 9* will see the second surface EVA – the first by the second pair of astronauts, with the original pair acting as IV crew-members as their colleagues experience the thrill of walking on the Red Planet. They would probably still be limited to a 2–4-hour excursion, and would complete the programme of deployment of equipment and local exploration begun during the first walk, but with probably less ceremony and more science. Post-EVA activities would follow the same pattern, setting the routine for ending each period outside.

*Days 10 and 12* EVAs will be performed by EVA Crew 1, and those on Days 11 and 13 by EVA Crew 2 – still in the local area of the lander, and probably on foot, to finally establish the lander area as a semi-permanent base of operations. Day 14 would be the second rest day, during which extensive medical examinations would be completed, with finalisation of the first traverse EVAs on the MRV and the decision about the feasibility of gradually increasing the duration of EVAs up to 8–10 hours over the third week, depending on real-time data.

*Week 3 operations (surface days 15–21)* By rotating EVAs on the surface and using the MRV, the first limited traverse in one of the sectors would be attempted, using drilling and sampling techniques further from the lander but not so far as to prevent a safe walk back in case of MRV breakdown. Depending on the activities and progress of EVAs on Days 15 and 16, exploration in the first sector would probably be extended over the following three days, with the science experiment network expanded to include remote stations to the limit of Sector 1, linked to the main science station at the lander area. Day 20 would be spent in the lander, with refresher training and simulations of ascent and rendezvous procedures, medical examinations, and analysis of samples. Day 21 would be the third rest day, and would also be used for planning and operational decisions for Week 4.

*Week 4 operations (surface days 22–28)* During this week the crew would move to Sector 2, essentially duplicating the activities and timelines of Week 3, with new areographical targets. The last two days of this week would again be occupied with training, simulations, medical examinations, sample analysis, planning and rest.

*Week 5 operations (surface days 29–35)* Exploration of Sector 3 would be completed, following the previous pattern.

*Week 6 operations (surface days 36–42)* The final Sector 4 exploration would be conducted, and while the last two days would again follow the familiar pattern, they could also be used as contingency EVA days for completion of any open issues prior to departure. In these final ten days on the surface, primary consideration would also be given to revisiting sites, installing permanent science packages, retrieving experiments and samples, and completing deep-core and heat-flow experiments – the final collection of data. All the geological sampling sites would be closed off during the week (subject to revisits), samples would be stowed for ascent, and data cassettes (or discs) and materials would be stowed for the long flight home. The MRV would be reconfigured for viewing the ascent and for a possible traverse to the second landing area. The final EVA close-out would certainly include some ceremony, marking the historical importance of the first period of human exploration of Mars.

The study team – in consultation with Apollo astronauts experienced on the lunar surface – strongly recommended that in addition to following the timeline for each EVA as developed in conjunction with the mission planners and orbital mission control group, the landing crew should also be allowed to take advantage of real-time discoveries and be given time to examine surface samples and features and to photodocument the EVA area. Due to the remoteness of Mars from Earth, a certain amount of flexibility will be necessary (as with Russian space station operations and activities on the ISS), rather than a minute-by-minute EVA timetable (as with Apollo). This would provide archival records for post-mission debriefings (still months away) and for updating crew-training protocols for Mars Expedition 2 and beyond.

*Pre-launch operations (surface days 43–45)* Depending on the final design of the lander or habitat and the mission profile, the final days on the surface would be allocated for stowage and for configuration for ascent. Landing support systems would also be isolated or mothballed. Some elements might need to be left for use on a later landing mission to expand operations as a revisit site (which was not achievable for the Apollo missions). After a pre-launch communication session that will certainly include some ceremony marking the impending first departure of humans from the surface of Mars, the departure would take place, after more than 1,000 hours on the surface, on Day 45.

*Orbital operations and the second landing* The four crew-members of the first lander, back with their colleagues, would begin a period of medical readaptation to microgravity, further analysis of samples, and interpretation of data. There would also be a long programme of debriefings to update the planning of the second landing. Exactly where and when the second crew would depart for landing on Mars would depend very much on the results, condition, experiences and observations of the first crew, and the condition of the surface at the chosen second landing area. The profile and protocol of surface activities at the second site would probably follow that of the first team, but with less ceremony and more science. This would provide a suitable and reliable alternative set of data for comparison with the work patterns and productivity of the first team. The first team of four would serve as the orbital mission control team for the second landing. With the completion and successful return of the second lander to Mars orbit, the combined crew of eight astronauts would complete their orbital activities (including the second landing crew's period of readaptation to microgravity), and prepare for the long journey home. After the six Apollo landing missions, two members of each crew were to be forever known as Moonwalkers. With one mission to Mars including two four-person landings and all crew-members performing surface EVAs, they could all could look forward to being forever known as Marswalkers.

## EVA ON MARS: THE OPERATIONAL ISSUES

The 1989 study was one of the first to seriously evaluate the potential of human activity on Mars.[7] From this study, NASA moved on to the 1993–94 Reference Mission study, which generated a further series of workshops and papers that discussed aspects of initial activity in greater depth and detail prior to the 2004 Presidential announcement of a return to the Moon and possible human flight to Mars. Using these latter studies as guidelines, together with interviews with leading specialists in field geology, robotics and Mars mission planning, a wider discussion can be presented on procedures during the first EVAs on Mars, and (in the next chapter) issues concerning the scientific research that would be conducted on the surface.

Marswalk One will probably be a ceremonial 'flags and footprints' traverse. This is how Apollo 11 – the first Moon landing – is depicted by its denigrators (and even by its supporters when discussing different approaches to exploration). Some remember the Moon landings as being only symbolic of superpower politics and political posturing, rather than the beginning of humanity's personal exploration of the Solar System. But the exploration of Mars will have to be different from lunar exploration, simply because of the distance, the time required for the journey, and the sustained health and wellbeing of the crew. Unless a contingency or emergency situation arises necessitating a premature return, it will be impossible to justify a mission to Mars for just another 'giant leap'. The scenario above outlines a potential mission profile, but very little about the early excursions on Mars is likely to be set in stone, and daily reviews and briefings will probably change at least some of what was planned for the next excursion. There are numerous other factors that will provide a more fluid approach for Mars EVAs.

### Early extended expeditions
Marswalks will begin with the first landing, and thereafter expand outwards. A Marswalk may begin like the first Moonwalks – on foot, for a short distance away

from the landing craft. There may also be longer traverses using an unpressurised rover vehicle – just an hour or so away from the landing craft. After that, a pressurised rover vehicle may be used for longer traverses away from the landing craft, but for regional rather than global trips, and perhaps for establishing a small encampment somewhere in the same region before returning to the landing craft. Such trips will only last days or weeks, rather than months.[8]

The NASA Reference Mission summarised proposed early EVA operations on Mars to support the ability of the suited crew-members to observe and explore the environment around them.

Since field geology is essentially involved with examining rocks and structure *in situ*, it is imperative that the suit visors be free of optical distortion and have as wide a field of view as possible, allowing the wearer to distinguish variations in colour and to discriminate between similar rock. In addition, any suit used on Mars will need to provide as much mobility as possible – overall personal mobility, and operational mobility for the length of the EVA, repetitive use, ease of maintenance, and repair and rescue scenarios. When full mobility is impossible, a suite of tools and aids should be provided for retrieving samples and equipment. All of these tools and hardware must be compatible with the environment in which they are to operate, and although maintenance levels during EVA should be considered, it should be minimal. Communication between the team on EVA and the support team in the rover or habitat, and navigation during the EVA, will form an integral part of the field geological activities, and should be integrated in terms of both operational and safety issues as well as research and science objectives.

In the 1997 Mars Surface Mission Workshop, with the preliminary guidelines from the Reference Mission, the team utilised a Mars outpost design as a focal point for basing EVA operations. This was an inflatable lab/hab structure near (or connected to) the surface lander, to be used for the duration of the surface activities. In reviewing operations on more than eighty Shuttle missions, the teams established a division of crew hours for each 24 hours spent on the surface as:

- Hygiene, meals, conferencing, maintenance, housekeeping and training – 7 hours a day.
- Pre-sleep, sleep and post-sleep activities – 10 hours.
- EVA, exploration, science, sample analysis, teleoperations and planning – 7 hours.

There was also concern – based on the experience of the longer missions on Russian space stations (especially a decade of Mir operations) – that the amount of crew-time spent on maintenance and housekeeping would be much greater as the mission increases and the operational life of the vehicle extends. It was suggested that a daily pattern should be incorporated to allow for a single-shift operation, with one or two days off over a 'weekend' and with crew recreational activities included in the timeline.

The workshop then suggested a six-step process of establishing a 'base' at the landing area in order to conduct operations while on the surface. These were:

- Deployment of power systems. In this scenario the power system would be deployed robotically via a separate cargo lander, and possibly repositioned 1–2 km from the landing site, where radiation hazards to the crew would be minimal. The astronauts would have to connect the system to their lander or supply during the stay on the surface, which might necessitate running electrical lines from the source to the habitat, depending on the chosen system.
- Deployment of a communication system to improve surface coverage and establish firmer links with the orbital crew (if there is one) and Earth.
- Deployment of In Situ Resources Utilization (ISRU) – probably robotically until the crew arrives. They would then have to check and test the ISRU systems, and might decide to move the facility to increase its effectiveness and expand its operation, or relocate it further from the ascent vehicle.
- Setting up the laboratory and/or habitable structure. This could be part of the design of the lander or be elements sent unmanned prior to the arrival of the crew, or could include an inflatable 'greenhouse'. Whatever the method, the habitation module would have to be reconfigured for operational use, as would the laboratory module to support science activities.
- Deployment of science rovers, probably initially by teleoperations as the crew adjusts to the ⅓ gravity and assess the surface conditions immediately after landing. There will undoubtedly also be unpressurised (Apollo LRV-style) local area rovers, and at least one large pressurised rover (mobile laboratory) for longer excursions, which may or may not be capable of telerobotic operations.

A landscape with no vegetation but covered in volcanic rocks and dust. It is freezing cold but there are billows of steam from geothermal activity. This is the interior of Iceland, but it is similar to conditions on the surface of Mars. (Photograph by Andrew Salmon.)

- The relocation of surface elements planned to land in a specific area might be difficult to achieve. It is highly conceivable that the crew might land further away from or too close to any unmanned landers that arrived earlier, and this might force a period of relocation to utilise the resources more efficiently and safely.

### EVA planning

Apollo EVAs were timelined to the minute. The astronauts had a list of tasks to carry out, and the timeline would be changed only if there were problems. In contrast, Marswalks will follow procedures more like the planning utilised onboard space stations, for which someone possesses just the basic skills to live in space,[9] and a person's experience and training is relied upon. This is particularly true for the geological traverses. Most of the science that will be conducted on the first Mars missions is discussed in the next chapter, but here it is worth mentioning some aspects that will possibly influence a given EVA or the planning for subsequent EVAs.

The NASA Reference Mission identified a range of EVA tasks that would be required to obtain the preferred scientific data; and as the surface mission matured, the EVAs would be modified. Although an Apollo 11-style EVA would probably be observed for the first few hours of the first EVA, the nature of the mission to Mars will involve longer and more frequent surface operations after the first landing. As a result, some of the early guidelines for EVA planning listed in the Reference Mission were:

- The 'buddy' system of paired EVA crew-members will always be used.
- During the encountering or handling of any object or sample on the surface, standard EVA protocol (gloved-hand access, no sharp edges, touch temperatures, simplified tool interfaces, and so on) must be observed.
- A safe haven should be available during all EVAs beyond walk-back distance.
- Seasonal effects (number of daylight hours, dust-storms, radiation events) must be taken into account when planning, timing and supporting all EVAs.
- The time-delay between Earth and Mars requires primary EVA support to be provided by the habitat crew. Orbiter personnel (if available) and Earth-based controllers can act as back-up, but real-time voice, video and data-relay between the EVA crew and lander is a priority. The loss of these links, could, depending on circumstances, terminate the EVA.
- Only one pair of astronauts should be outside the habitat or pressurised rover at one time, although two pairs may be outside in extreme cases – such as for maintenance and/or repair, or for one pair to rescue the other pair.
- EVA during the martian night will be trained for and certainly possible, but would probably not be part of nominal planning and would most probably involve contingency EVAs local to the pressurised habitat or rover.
- EVA suits must have minimal pre-breathe and minimal turnaround between uses, for both operational planning and emergency situations.

The Mars Surface Mission Workshop also identified several areas in which continual activity will affect the planning of EVAs. These included walkaround and inspection of the facilities, spacesuit maintenance and cleaning, TV surveys of the facility by robotic rovers, change-out of external elements such as filters, and the maintenance and repair of internal elements. In order to achieve this, there would need to be provision for a workshop or machine-shop area. As the crew becomes more accustomed to working in the martian environment, it might be possible to cannibalise and rework items, or even manufacture replacement hardware.

Mission rules are already in place for a dual-person EVA, although for historical, publicity and logistics purposes, probably only one person would access the surface for the first few minutes of the initial EVA, to lay claim to being the first person to walk on Mars, even though they will have landed with several other astronauts. Most reports indicate two pairs for field-crew operations, with one pair being on the surface at any time and the other team working in the laboratory or serving as EVA controllers in the lander. The IVA crew would also be operating unmanned rovers, performing maintenance work, and following the work of their colleagues on the surface.

Initially, the first EVAs will be walking traverses a few kilometres from the lander – planned radially to record observations, take digital photographs, and collect a few samples. Data from these short EVAs will be analysed in the lander in order to plan more distant traverses supported by unpressurised rovers.

A programme of intermediate EVAs of no more than a working day's operation (up to 10 hours) could then follow. These second-phase excursions would be dependent upon lighting and surface conditions, and would include systematic coverage of the landing radius out to interesting features and targets of opportunity (similar to proposals in the 1989 study) – possibly as simple as points of a clock face. Surface samples would be collected and cores would be obtained, and there would be provision – should the area be of sufficient interest – for return visits. This intermediate phase would also include deep drilling operations and traverse areophysics. At the same time, the unmanned rovers could be utilised to push the exploration boundaries further out (before 'risking' human explorers), or to explore regions beyond the safe limits of the crew, such as deep ravines, cliff faces, and steep slopes. They could also provide close-up photographs of surface features and textures. Information from these EVAs will have to be carefully analysed in the laboratory, in order to effectively plan the next phase of the exploration. This may require periods duing which no EVAs are conducted, thus allowing time to evaluate the data so far gathered, repair and maintain hardware, rest the crews, and update the objectives.

After all the one-day EVA targets have been reached, there should also be a period of reflection to allow both the crew on the surface and Earth-bound scientists to evaluate what has been learned. Based on these data, long-duration excursions could be conducted using pressurised rovers and travelling far from the habitat. The targets for these extended EVA operations will already be known from planning the overall mission (primary and secondary objectives), but would probably not be confirmed until the exact landing site has been determined and the results of the early and intermediate EVAs have been analysed. The EVAs in the intermediate programme would be expected to take up approximately 25% of the total surface time (125 days for a 500-day stay) before extended operations away from the base for several days, or, dependent upon capability and the real-time situation, several weeks.

Depending on the final mission scenario, and situations on the surface when the first astronauts arrive at Mars, after the initial local EVAs and a period of intermediate traverses local to the lander there should be a programme of extended EVAs that could be mounted by part of the crew for several days or perhaps a couple of weeks. These extended EVAs may take 50% of the crew away from the lander, and so normal activities at the lander would have to be continued with half the crew, and local EVAs would probably still have to be conducted for housekeeping and contingency situations. Between these long traverses, the intermediate EVA sites will also probably be revisited when and where the opportunity arises, to gather more data or confirm earlier findings.

Extended EVA operations will probably require strategic planning, including outposts where sources of fuel, water, oxygen and spares could be sited, safe havens, contingency or emergency protection, and early warning and rescue facilities. Unmanned robotic devices (rovers, aerobots, remote sensors) will support and supplement data from the human activities, and this network will probably overlap to allow walk-back contingency, automated rover 'rescue' traverses, or far greater EVA operations than the local or intermediate EVAs could safely and efficiently support.

In any phase of a spaceflight, safety procedures must be a priority – and this will be particularly true on extended EVA operations on Mars. In the design of hardware, procedures and operations, safety and operational limits will be incorporated in the design of the equipment, and the terrain and conditions on Mars will also result in 'go/no go' decisions. Such constraints will include (among other factors yet to be determined) malfunctioning EVA suits, problems with the manned rovers, malfunctions or loss of communication between the field crew and those on the habitat, major system failures of EVA hardware, the habitat, ascent vehicle or orbital spacecraft, immobilisation due to dust-storms (which could be local or global, lasting weeks or months), degradation of the temporary habitat or facilities, the medical condition of the crew, and major solar flare activity.

Recent studies have indicated that a pair of two-person rovers will be carried on the mission – one used for the EVAs, and one for reserve and rescue facilities, capable of returning four astronauts over a short distance. Some reports have indicated the use of two separate rovers with towing capability in the event of a breakdown; but the use of more than one pressurised rover on the first mission is uncertain, and is probably not an option for a one-lander mission, due to mass limitations. However, if these pressurised rovers can first be landed unmanned (as was planned for extended surface operations during the Apollo Applications Program), then perhaps a two-vehicle expedition could be mounted with more reliability and safety – particularly if smaller unpressurised crew rovers or flying vehicles are also available.

Such an EVA infrastructure on Mars would probably support a second visit to the area, or one adjacent to it, in order to capitalise on the hardware left behind from the earlier crew, and for greater use of resources and improved safety parameters.

### Communications

Current human mission operations are operator-intensive, but the expected complexity of local mission operations planned for a Mars mission could render attended monitoring and control techniques unaffordable. Because of the long round-trip transmission time, many real-time operational decisions must be made at Mars. Such local decision-making must be assisted by expert (computer) systems, and the links for distributing information rapidly to local terminals must operate largely unattended.[10]

During Apollo EVAs, the astronauts had immediate communication with the CapCom and the geologists at Mission Control. This would not be feasible on Mars, and communications with Earth would include a debriefing at the end of each day. A comparison would be the Mir or ISS science scenario, in which astronauts with the essential skills work to a personal regime and report their progress at the end of each day.

During a traverse on Mars, direct communication with Earth is improbable, but communication with the landing craft would be routine, and the astronauts on the surface would therefore work independently.

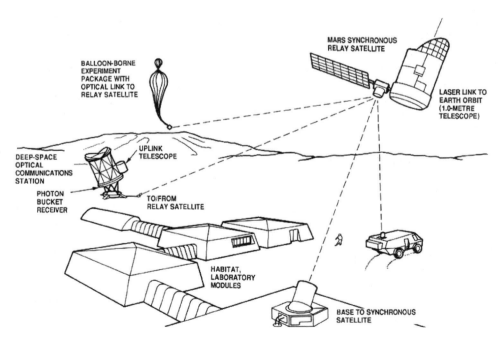

A JPL concept of a fully operational Mars base with an extensive communications network to all parts of the mission as well as back to Earth. (Courtesy NASA/JPL/ British Interplanetary Society.)

## EVA communications

As with all EVAs, communications will be essential on Mars to monitor the progress of traverses and the safety of the astronauts on the surface or in the rover. EVA suits will incorporate a range of communications aids, including a heads-up display in the helmet, and a helmet-, wrist- or arm-borne recorder for vocal comments. The data would be passed to the laboratory module on Mars, and communications would be regularly updated with the mother-craft in orbit. Once each day it would be transmitted to Earth via the dish on the mother-craft. The martian weather will interfere with EVA communications between astronauts and to the rover or to the laboratory/lander/habitation modules. The effects are analogous to those produced by dust- or sand-storms on Earth: static electricity and radio interference. During EVA, a high-data-rate communications link could be used to show the laboratory crew what was being seen.

A central station in the lander or habitation module would control links to local and remote astronauts. Channels could be automatically assigned as required, and sized in bandwidth, power and processing as appropriate. The network would automatically detect and analyse malfunctions, and appropriate action could be taken either automatically or manually via alternative communication nodes.

An astronaut may make a service request by voice or by keypad. The channel, function and bandwidth would result in a routing via direct line-of-sight radio frequency transmission where possible, or a console operator in one of the modules may send non-critical commands to remotely operated sensors on the EVA suit. Commands might also be generated by the astronaut, with head or eye movements, or even cortical activity.

The heads-up display, microphone and speakers in the helmet would provide video, text, and suit and life support data, and telerobotic operations would be made possible by tactile/force feedback through the gloves, synthetic speech and audio through the speakers, and video, graphics and alphanumeric displays on the heads-up display.[11]

The development of effective EVA communications is a priority. In most reports, local facilities for communicating with the lander or a satellite network have been proposed, but in the 1999 paper by the Cornell Team[12] it was suggested that a 'start-from-scratch' approach would be adapted on the first EVAs until an expanded system could be developed. Assuming as the target a foot or rover traverse of up to 8 hours, a 30-km range for a surface communication system would be adequate, if the astronauts travelled at 7 km/h for 4 hours out from the lander and at the same speed and over the same time on the return. Allowing for station stops and pre- and post-drive activities, cost-saving and mass-saving alone would support a ground-based EVA communications system for local and intermediate EVAs. In the report, the team reviewed four types of system and their effectiveness for a primary surface mission. The following compares the communications systems:

| Communications network | Infrared system | Fibre optics system | Satellite system | Reconfigurable wireless network |
|---|---|---|---|---|
| Support navigation | Yes | – | Yes | Yes |
| Mobile/flexible | Yes | – | Yes | Yes |
| Robust | Yes | – | Yes | Yes |
| Easily repairable | Yes | – | – | Yes |
| Practical set-up | – | – | – | Yes |
| Upgrade/extendable | – | – | Yes | Yes |
| Flight-tested | – | Not large-scale | Yes | – |
| Maximum range | – | 4 km* | 3,900 km | 30 km |
| Network mass | – | 175 kg/km | ~8,000 kg | ~180 kg |
| Power requirement | – | – | Solar/battery | 10 W (EMU) |
| Mars dust factor | Not good | Not good | – | Good |

* 100-fibre single-mode loose-tube cable without splicing

The study suggested that wireless radio was the most mass-efficient and cost-efficient system for the primary mission, and that reduction of the carrier signal and its range would be limited to the surrounding terrain. It also discussed the use of relay communications through base stations (similar to mobile telephone technology on Earth), but this would require setting up, which, in addition to the other EVA tasks and the routine base activities, would complicate telerobotics operations. Such remote stations might be an objective for intermediate EVA operations, overlapping the basic range by triangulation. Experience from military operations and activities in Antarctica has demonstrated that these systems require a robust communications network due to surface conditions, including the ruggedness of the terrain, and extreme isolation. If a large natural obstacle blocks line-of-sight communication, repeater facilities could be erected to 'bend' signals around it. Extended EVA operation far from the base will probably have to rely on satellite communication technology (like the GPS system on Earth, or the TDRS system used to support Shuttle operations). These communications satellites (10–1,000kg in mass) could be sent from Earth to Mars orbit prior to the crew's departure, and could be supplemented by deployment from the parent craft prior to landing, based on the status of the communication system and upgrades to the flight plan.

**Field camps**
The study suggested sample payloads and associated mass values that might be considered for remote field camps offering semi-permanent residence and storage facilities in extended-duration activities on the surface. These would initially be based within the limit of a walk or a trip on a rover, and would overlap the radii of early operations, thus extending the operational range of the crew and vehicle while working within safety guidelines. These field camps could also be used for communication relays, remote data-collection and storage of consumables, or as (perhaps inflatable) facilities for rescue, vehicle repair and maintenance, and EMU repair and servicing.

They could also be used as a location for dedicated equipment (nuclear power plants), stores (sample payloads and associated equipment) or scientific research (deep drilling hardware), and would alleviate the need to carry items to and from the site. Payloads deposited at this location could include equipment for field geology (hand tools, sample containers and documentation equipment – 335 kg), instruments used for traverse areophysics (400kg), sets of areophysical and meteorological instruments for surface emplacement (200kg), a 10-metre drill (260 kg), and a 1-km drill (20,000 kg).

**Field geology**
Samples collected during field expeditions might include meteoric pieces of Earth (and possibly Venus), and specimens of the martian mantle excavated by large impacts.[13] Incidental specimens will include evidence of cosmic rays and solar activity[14] (although rocks from the surface of the Moon are more productive, due to the lack of a lunar atmosphere). Sample selection will often be based on experience, and it will not be determined whether a sample is the type required until it is taken to the landing craft and analysed. However, career astronauts with sufficient knowledge of, or experience in, field geology may have evolved an instinct for such work. The results of sample analysis would be instrumental in refining later field

An earth sample of haematite. Mars has areas of anomalously high mineral content, such as the haematite at Terra Meridiani. This can be formed in areas of geothermal heating where hot water dissolves iron-bearing minerals. It may also be laid down by transient lakes or when soil is briefly moistened by water. (Courtesy Lapworth Museum of Geology, University of Birmingham.)

expeditions. Due to mass limitations, not all specimens could be returned to Earth, and superfluous specimens would be discarded; but the sites producing outstanding specimens would be revisited to obtain further or even better samples. The best samples would be returned to Earth, but the abilities of the astronauts on site will have a major influence over the planning of the next traverses.

The skills of field geology are acquired by experience. However, advance preparations – such as geological maps, and an organised approach – are necessary for a field-trip. There are two aspects to this:

- Observation of the terrain and environs – which appears completely different on the ground than when seen from orbit. (On Apollo 15, Dave Scott looked out of the top of the Lunar Module to observe the surroundings.)[15]
- The collection of samples. Field geologists do not rely on pictures taken from orbit, and on the surface they need only a hammer.

According to field geologist Graham Worton: 'Everything we do in field geology relates to a place, an incline or a layout.' The astronauts could leave the landing craft to explore the region, but they would need to be able to find their way back, and to note the location of sites of interest for reinvestigation. The number of samples that they could carry would be limited – even with a robotic assistant or a rover. On Earth, GPS Navstar navigation satellite systems are routinely used, and there will probably be a similar system in Mars orbit. This is a trivial expense compared to the overall cost of the mission.

## EVA SUPPORT

The question arises: 'Why send humans to Mars when samples can be collected by robots?' The primary answer is that field geologists have instinct, experience and skill, and traverses can be planned day by day, based on the results of analysis with the tools available in the field or in the laboratory.

NASA's 1980s plans for a sample-return mission incorporated robotic rovers to collect samples and bring them to a central point for automatic loading onto a Mars Ascent Vehicle and return to Earth.[16] However, an exclusively robotic sample-return mission would need to gather possibly hundreds of samples to equal the mass and variety collected by, for example, the astronauts of Apollo 15, whose geological skills were secondary to other requirements for the mission.[17]

### Robots on surface traverses
In supporting a human mission to Mars, the role of robots would include the following:[18]

*Field-work* Robots would allow humans to project their intellect far beyond their base of operations. Robotic sensors can operate at wavelengths other than visible light (infrared and ultraviolet, for example) and robots can survey local sites and gather geological samples for study at the base – particularly if conditions are too

Support for human activation, which might include automated return of samples.
(Courtesy Pat Rawlings.)

hazardous for the astronauts to venture out. Telepresence would also be used as an option to having a geologist on site. Robots are also particularly good at reconnaissance, gathering large amounts of data, and carrying out simple analysis, all of which are repetitive. They are relatively difficult to reprogramme for new tasks, but would be predictable, and could be directed to test hypotheses suggested by the data that they have gathered.

*Detailed surveys* Robots would act as scouts. This might consist of an orbiting craft able to provide high-resolution images of surface features – in much the same way that spy satellites can produce imagery for direct transmission to terminals at headquarters. Areas of interest could be closely examined by geologists.

*Maintenance and logistics* Robots could replace humans by conducting routine and mundane tasks, and could also provide emergency assistance to humans.

*Difficult or dangerous terrain* Robots would act as surrogates for the human explorers. Astronauts in a pressurised rover or in the habitation module at the landing site would be able to control rovers roaming either in the immediate location, or even far away, wherever the terrain was unsuitable for human EVAs.

They may even be sent to Mars ahead of the human landing, to be controlled either from Mars orbit, from Phobos, or from the landing site. Astronauts will probably land near the equator and then control robots to bring material back to the habitation module. If robots are sent in advance they could stockpile samples at the landing site.

### Partners working together

Humans and robots can work as partners. We may choose to indefinitely defer sending humans to Mars and instead focus on the development of sophisticated automation and robotics, but the best solution is to send both humans and machines to work as interactive partners. Humans are necessary for geological field studies and the search for signs of extant or fossil life (using geology and biology in the specialised field of exobiology). These are areas in which humans have broad experience and the ability to link unexpected and disparate observations in the field. The humans would need to be relatively safe and comfortable, and with appropriate use of automation and robotics they would be productive. Around the world, humans work with robots in numerous different ways. NASA, however, might be thought too prescriptive in its approach to using robots as 'glorified shopping trolleys',[19] and it would be more appropriate for astronauts in a habitation module to control them at a long distance or in difficult terrain. Robots on Mars cannot be controlled from Earth in real time, but humans on Mars would be able to do so.

Humans and robots working as a partnership may be essential in Mars gravity compared to the $\frac{1}{6}$ -g lunar surface. EVA suits on Mars will feel heavier, and more equipment will be carried on the field trips, and so NASA's 'shopping trolley' robots may still play some role. Equipment on Mars will weigh much more than it would on the Moon. The Shuttle Extravehicular Mobility Units (EMUs), for example, are ideal in microgravity, but on Mars they would be too heavy. Moreover, a suit carries an array of equipment for life support and protection, and there is little capacity for extra weight. The robots would be used like pack animals, similar to the Apollo 14 Modularised Equipment Transporter.

Equipment taken to Mars will need to be compact, and not as state-of-the-art as comparable equipment being used on Earth at the time of the Mars landing. There will be a limit to the power available for the instruments, a limited level of technology, and a limit to the data and samples that can be returned to Earth. Power on Mars can be provided by the Sun (as used by Mars rovers so far), or by a nuclear reactor or a radio isotope thermal generator (in which plutonium emits heat for operating either thermocouples or stirling engines, which convert movement into electrical power). It will be necessary to have very-high-bandwidth communications – a broadband link from Mars to Earth encompassing the full spectrum of audio/visual data.

Complex state-of-the-art equipment could be carried on the rover, but the failure of such equipment would be more difficult to remedy, as it could only be repaired by humans with expertise in computing, electronics and mechanics.[19] Conversely, simpler instrumentation would provide assurance in returning information to the habitation module or to Earth.

## THE DANGERS OF EVA

Contamination cannot be avoided. Suits leak oxygen, and (like ordinary clothes) shed fibre particles and microbes, which could confuse scientific readings. Analytical instruments on Earth are now so sensitive that even the merest trace of bacteria can be amplified and detected.

Samples could be stored 'in the open', or sealable bags could be used,[20] but viruses and bacteria – the 'calling card' of a human – must not come into contact with the samples. The search for extant life on Mars would be improved by having humans on the planet; but paradoxically, their presence will complicate the search.[11]

As part of the Planetary Protection regime, all equipment and vehicles intended for Mars will need to be sterilised – a procedure followed for most (if not all) American and Soviet robotic landers. Sterilisation involves either dry heat or gamma-irradiation – both of which penetrate deep inside components – or treatment with chemicals such as ethylene oxide.

During Apollo there was no talk of forward contamination,[20] but the study of martian samples will be very different. It was always considered very improbable that lunar regolith or rocks would contain life, but on Mars there is a possibility of past or present life on or near the surface. A considerable amount was learned about forward contamination from research in the Antarctic. The early researchers left a large amount of debris – especially at the bases – but this has now been removed. Outside sanitation is no longer allowed in the Antarctic. An enclosed facility must be used, and waste must be taken away. In the same manner, waste produced on Mars will have to be taken to the base and sent back to the parent craft, from where it could be placed in long-term orbit, fired towards the Sun (although this would require a large amount of energy), or brought back to Earth. For a long stay of, say, a year, greenhouses using human waste will probably be used.

The astronauts will also have to guard against back-contamination, or even just the carriage of martian life without being aware of it. The crew will require full protection when working with samples in the laboratory. They will not work with ungloved hands, but will use a glove-box to prevent being contaminated by any toxic chemicals or native life.

At present, all known life exists on Earth – but it could be that billions of years ago an impact on Earth sent matter, carrying life, to Mars. The search for life can be best carried out by field geologists with the appropriate knowledge, skills, instincts and experience – and robots do not have these facilities. If life *is* found to exist on Mars, it will have to be determined whether it is indigenous or whether it was transported by humans; but if none is found, the search strategy will need to be re-evaluated. Continuing attempts to find life will lead to changes in EVA procedures, and could even change the emphasis of the entire programme.

## HAZARDS ON MARS

On Mars there are many hazards which could affect the planning and execution of traverses. Some of these hazards – such as dangerous terrain, adverse weather, and radiation – are familiar, but others are unique to the planet.[21] Advance warning of solar activity and martian dust-storms would be a necessity. Minerals mined on Mars could be used for lining underground tunnels for protection against radiation, but there will need to be a shelter within convenient reach of the crew at all times, even if they are far from the lander. Statistically, however, landing and take-off carry the most risk.[20]

### Dust-storms

Dust-storms range in intensity from dust-devils, through regional dust-storms, to global dust-storms.[22] Artists' impressions of martian dust-storms often depict an almost solid wall of opaque dust advancing over the landscape – but they may be quite different. In the thin atmosphere, dust is not easily raised, and even the worst dust-storms have only the effect of a gentle breeze on Earth. For a human, a martian wind of 111 mph would be the same as a wind of 12 mph on Earth. (The Viking landers measured wind speeds of 30 m/sec (65 mph) at a pressure of 7 mbar.)

Mars is dominated by dust-devils a kilometre or more high,[23] and the dust that they stir up may be the seed for the dust-storms. They leave tracks as they roam the deserts, although their speeds have not been measured. Recent research suggests that they produce electrostatic effects leading to radio interference and problems with electrical and electronic equipment.[24] These electrostatic forces also pull dust-grains together to form aggregates – small balls of dust the size of sand-grains, which break apart on impact (and therefore do not cause abrasion).[25] On a larger scale, dust-storms may darken the sky to partly block out the Sun (as recorded by Viking lander images), and will also obscure solar panels. Very fine particles will enter machinery or spacesuits – even without a dust-storm. The dust may also be toxic, attacking plastics or affecting people in the same manner as silicosis.

Dust-storms are more common at certain times of year. The global events tend to begin during the southern hemisphere summer, when Mars is at its closest to the Sun. The global storm of 1971, for example, reached to a height of 70 km, and covered the 25-km-high Olympus Mons.[26] However, due to the thin atmosphere, the eroding effect on the landscape (and on spacesuits and vehicles) is far less pronounced than it would be on Earth. Anemometers could be used to measure wind speed and direction, and a lidar (laser–radar) could be used to measure the amount of dust in the atmosphere.

Martian dust-particles are only a few microns in size (much smaller than sand-grains), and after a storm, the dust remains suspended in the atmosphere for perhaps several months. (Red light is scattered, and the sky has been variously described as either salmon pink or yellow-brown.) This could have a serious impact on crews on the surface. If the dust is found to be harmful, or continues to be a hazard to the equipment (especially the spacesuits), it may be that subsequent traverses are conducted only by robotic rovers. If the problem is sufficiently serious, the entire landing might have to be abandoned.

Dust can be moved only by a storm, and very little particulate matter is seen in the air around a lander, even during gusts of wind. Wind-blown matter on Earth consists mostly of sand-sized grains of quartz, weathered out of the predominant rocks of Earth's crust. Quartz is tough, and does not erode into dust-sized particles, whereas martian wind-blown materials are dust-grain-sized particles of iron-rich clay, eroded from volcanic rocks.

### Radiation

There is a constant background of cosmic radiation from deep space, but a planetary atmosphere absorbs much of this radiation and shields explorers. It may be that a great danger arises from the effects of solar flares, which are sporadic. Flares cause problems which last for a day or for several days, and the astronauts will need to stay under cover for that length of time, whether below ground or in a vehicle or a facility. They would possibly use water bags to absorb the radiation; but the best material (recently developed) consists of large slabs of high-density polythene called POLY, which is not particularly heavy but is packed with hydrogen atoms which act as radiation moderators. Electromagnetic radiation takes about 12 minutes to travel from the Sun to Mars; but matter takes between an hour and a couple of days.

Mars has no global magnetic field, but the rocks preserve a 'fossil field' which existed four billion years ago,[27] and which was possibly as strong as Earth's current magnetic field. Some locations on Mars might even provide limited radiation shielding, which would provide an incentive to site bases there, or to at least have them within easy reach of EVA crews.

### Medical considerations

Normal background radiation on Earth has been recorded at 0.4 rem per year, with a 5-rem-per-year limit set for high-risk occupations. Up to 25 rem in a lifetime is not fatal, but in excess of 500 rem over the course of a human lifespan (70 years) could be lethal, depending on a variety of physical characteristics that may affect this figure by a factor of two. Although NASA has set limits for radiation exposure in low Earth orbit, no limits for a Mars mission have been specified (or at least, not published), although it has been suggested in published papers that a maximum dose of 100 rem would be probable for the entire mission, with 5–10 rem being accumulated during the EVAs. The following are the radiation exposure limits, in rem, for missions in low Earth orbit, as specified by NASA:

| Exposures interval | Blood-forming organs | Ocular lens | Skin |
|---|---|---|---|
| 30 days | 25 | 100 | 150 |
| 1 year | 50 | 200 | 300 |
| Career | 100–400 | 400 | 600 |

Primary factors effecting human performance were discussed during the Science and Human Exploration of Mars conference in 2001.[28] These factors included low alertness, fatigue, adaptation to the workplace, motivation, a healthy brain and mood, focused concentration, physical interface with the workplace, and a sensible workload. The habitat would need to be maintained, and a programme of housekeeping sustained. Exercise would be supplemental to surface activities, and there would need to be a programme of recreation and privacy. A crew healthcare programme – including protection from radiation – would also have to be established.

In the 1997 Mars Reference Mission, the most challenging scenario identified fifty-five risks, and 343 critical questions in twelve risk areas. In responding to these (and many more still to be defined), a Critical Path Roadmap became the blueprint for a focused effort in research and technology to reduce these risks or to prevent them entirely during human activity on Mars. In direct connection to surface operations, thirty-five risks and 233 critical questions were identified in three risk areas: habitation systems (including advanced life support systems), environmental health monitoring, and food and nutrition. Medical care systems focused on clinical capabilities and multi-system alterations, while adaptation and countermeasure systems included bone loss, cardiovascular alterations, human behaviour and personality, immunology, infection and haematology, muscle alterations, neurovestibular adaptation, and the effects of radiation.

In the Critical Path Roadmap, radiation effects and surface radiation environment – including adequate shielding – were of primary consideration. Environmental issues concerned dust and biohazards, and the prolonged effects of 0.38 g require further study. It is clear that during surface operations an effective programme of gravity augmentation exercises should be adopted using various exercise concepts – ergometer, treadmill, resistive devices, and so on.

In addition, there are the problems of behaviour and performance within a small group of multicultural, dual-gender, highly motivated individuals. Proper experience, training, teamwork and compatibility is essential for the safety and success of the mission. Particular concerns for the smooth harmony of a 'good crew' will be the extended duration, operating in such a remote location, the requirement for high autonomy, high-risk operations (in terms of expense and danger), and a high visibility, with the additional pressure to succeed. Measurements would be taken of sleep and circadian rhythms, physiological adaptation, and neuro-behaviour and interaction between the human crew and with the robotic devices, and there would be contact with Earth and 'home life' to determine the state of health of the crew. Coupled with medical examinations and studies of the crew-members' work performance as the mission proceeds and the objectives are met, the end goal would, of course, be for all of them to return to Earth happy, healthy, and with a sense of achievement.

## Mobility

A final factor in deciding what would be carried out during traverses is the mobility of the astronauts and vehicles. Apollo EVA planners particularly remember the time constraints. There was so much planned. When there were problems they would act differently, or simply pause; but they had to complete a task as best they could, and begin another task because they had run out of time. On Mars, if a scenario similar to the 1989 proposal were to be followed there would be more time to complete such tasks and amend the EVA programme accordingly. If there were vehicles available and the conditions made their usage feasible, this would make a huge difference both to the range of traverses and the potential area that could be covered. This would also encompass terrain that was never an option for Apollo exploration on the Moon – such as the bottom of craters or up the sides of high escarpments. Small unpressurised rovers would be used close to the landing craft, but pressurised rovers would be used for trips further out. Fliers considered for advanced Apollo missions would be an option on Mars, as there is a thin atmosphere, and so it would be possible to use aircraft with very long wings (similar to the U-2 spyplane). Another option would be to use variants of the Lunar Flying Vehicle or rocket back-packs.[29]

In the official reports it is envisaged that the crew will utilise small robotic rovers and mobility aids, as well as larger unpressurised and pressurised surface rovers, in their exploration of the surface. The design of the rovers will be dependent on further study, but it seems clear that a combination of Apollo LRV design and what was planned as MOLABs for later Apollo missions will feature in surface activity on Mars. In the 1997 Reference Mission report it was suggested that journeys of several hundred kilometres in pressurised rovers over several days or weeks could be mounted between resupply, operating a reduced-scale programme similar to that of the lander module, with a smaller telerobotic vehicle and a surface exploration programme. In the NASA Reference Mission that indicated their use, a craft would be landed unmanned prior to the arrival of the surface crew, due to increased mass and complexity. The rovers would be available for the first crew, but due to safety limitations, and to prove the concept for perhaps the second landing mission, they would be used in limited journeys close to the lander. The maximum range and duration of rover-supported operations will be dependent upon the type and number of vehicles available and the ability to maintain them in the field, either inside a pressurised hangar/workshop or on the surface. The unpressurised Mars Roving Vehicle (MRV) should be considered as an extension of the EVA suit. Dual operations with rovers would allow extended operations, and distant sites could be explored while operating a safety redundancy for the crew.

With such a wide range of factors to be taken into consideration – even just to step outside and become the next Neil Armstrong – it is easy to forget that any human mission to Mars would be a major scientific field-trip. After the euphoria of landing and planting the flags, the crew would be commited to a serious programme of scientific investigations.

## REFERENCES

1 Benton C. Clark, *Crew Activities, Science and Hazards of Manned Missions to Mars*, IAF-88-403.

2  Eric M. Jones and Harrison H. Schmitt, 'Pressure Suit Requirements for Moon and Mars EVAs', submitted to the American Society of Civil Engineers, Space 92, 31 May 1992, Denver, Colorado; LA-UR-91-3083 (revised).

3  Apollo 17 Technical Crew Debriefing, 4 January 1973, MSC-07631, Training Office, Crew Training and Simulation Division, NASA MSC, Houston, Texas.

4  'EVA in Mars Surface Exploration', Final Report, 31 May 1989, prepared for NASA, LBJ Space Center, Houston, under the Advanced EVA Systems Requirements Definition Study, (NAS9-17779, Phase III), prepared by a cooperative study team from Essex Corporation, Camus Incorporated and Lovelace Scientific Resource Inc.

5  'Generic EVA Tasks at Mars' 1989 Report, p. 31, Table 2-1.

6  NASA JSC hand-out, World Space Congress 2002, Houston, Texas.

7  Interviews with Graham Worton, 4 December 2003; Charles Frankel, 9 January 2002; Michael Duke, 28 January 2004; Bo Maxwell, 5 and 16 December 2003; Ian Crawford, 15 January 2004; Alex Ellery, 10 October 2003 and 4 February 2004.

8  Carol R. Stoker, *Science Strategy for the Human Exploration of Mars*, AAS 95-493.

9  Andrew Salmon, 'Mir: workshop and laboratory', in *A History of Mir, 1986–2000*.

10 *Report of the 90-day study on human exploration of the Moon and Mars*, NASA.

11 *Extravehicular Activity at a Lunar Base: Advanced Extravehicular Activity Systems Requirements Definition Study*, NAS9-17779.

12 *Extravehicular Activity Suit Systems Design: How to Walk, Talk and Breathe on Mars*, Cornell University.

13 Ian A. Crawford, 'The scientific case for renewed human activities on the Moon', *Space Policy* (in press).

14 Ian A. Crawford, 'Back to the Moon?' *Astronomy and Geophysics*, **44**, 4.

15 David M. Harland, *Exploring the Moon*.

16 *Mars Rover: Sample Return Mission Study*, JPL/JSC/SAIC, AAS 87-195.

17 Interview with Ian Crawford, 15 January 2004.

18 *Exploring the Moon and Mars: Choices for the Nation*, OTA-ISC-502, 1991.

19 Interviews with Alex Ellery, 10 October 2003 and 4 February 2004.

20 Interview with Mike Duke, 28 January 2004.

21 *Safe on Mars*, National Academy Press, 2002.

22 R. J. McKim, 'Telescopic Martian Dust Storms: A Narrative and Catalogue', *Memoirs of the British Astronomical Association*, **64**, 1999.

23 Intermarsnet Phase A study, ESA SCI 96 (2).

24 'Electric and magnetic signatures of dust-devils from the 2000–2001 MATADOR desert tests', *Journal of Geophysical Research*, **109**.

25 Joseph M. Boyce, *The Smithsonian Book of Mars*.

26 Andrew Wilson, *Solar System Log*.

27 *Science*, **279**, issue 5357, p. 1976–1980.

28 John B. Charles and Thomas A. Sullivan, *Science and the Human Exploration of Mars: Risks to the Crew on the Surface*, Bioastronautics Office, NASA JSC.

29 David J. Shayler, *Apollo: the Lost and Forgotten Missions*, Springer–Praxis, 2002.

# Science on the surface

The majority of studies and articles about human missions to Mars have, understandably, concentrated on the daunting (and risky) journey to the planet; but far less has been written about what humans will do when they arrive there.[1] To some degree the journey will be an epic achievement in its own right, and so the omission is understandable; but it is unfortunate that the next stage in the annals of exploration has been neglected. The human exploration of Mars could so easily follow the scenario of the Apollo missions: flags and footprints, followed by the real science – by which time the political paymasters and the general public have lost interest. Hopefully, by the time of the first human mission to Mars there will be a new priority for public expenditure, possibly under different political leaders with a greater interest in science.

## AREAS OF RESEARCH

When humans finally reach the alluring Red Planet, the research will encompass three primary areas: geology, atmosphere and life.[2] It has even been argued that life – past, present or future – will be the ultimate focus.[3] This would encompass the search for fossils or biomarkers, uncovering signs of present-day life, and determining whether indigenous resources can be used to support expeditions or outposts, laying the groundwork for future colonies.

The scientific disciplines that will be followed on Mars are more varied than those pursued on the Moon – because Mars is a planet, much more like the Earth than the Moon, and it has its own complex history which we are only now beginning to piece together. Scientific investigations would include the following:[4]

- Interior, crustal structure and activity.
- Geochemistry and petrology.
- Stratigraphy and chronology.
- Surface processes and geomorphology.
- Ground ice, ground-water and hydrology.

An Earth sample of olivine basalt. Low-latitude dark regions on Mars, such as Syrtis Major, are found to have the mineral olivine present. Olivine soon breaks down in water. (Courtesy Lapworth Museum of Geology, University of Birmingham.)

- Life, fossils and reduced carbon.
- Lower atmosphere and meteorology.
- Climate change.
- Upper atmosphere, ionosphere and solar wind interactions.

In essence, these disciplines cover the history of the planet[5] – the story of its formation, its geological record, and its climate (which on Earth would be the ice ages and the greenhouse effect).

### A reason to go

Is science a justifiable reason for sending humans to Mars? Michael Duke and many of his colleagues agree that science is not the principal reason for going,[6] and that there must be other (higher-level) considerations. The Apollo missions were spurred on due to competition with the Soviet Union. This was Cold War political rivalry and a possibly unique set of circumstances, and science was a secondary consideration. Whatever the primary reason for going to Mars, science may be instrumental in the decision.

### Tools of the trade

The investigations to be carried out will depend on the mass of the equipment, the availability of power supplies, where it is to be situated, and whether it needs to be maintained by a crew. There is an argument for keeping the equipment as simple as possible,[7] but it will need to be robust to withstand the martian weather. Even on a human mission, in the event of a malfunction there will be no chance of obtaining spares unless they are intentionally carried. The question of whether the equipment needs to be crew-tended leads to whether humans are needed at all; but as already mentioned, field geology is best carried out by humans.[8] The possibility of conducting field science using artificial intelligence and robots appears remote.[9]

Equipment could be deployed – either by robots or by humans – for data to be collected and relayed to Earth for long-term study (analogous to the ALSEP stations deployed by Apollo astronauts on the Moon). It could be close to the landing site, over a region as part of a network, or across the planet in a global network (as planned for the JPL MESUR robotic network on Mars in the 1980s). It could also be carried on rovers (possibly teleoperated by humans sited on a martian moon or in orbit around Mars), or placed close to a future landing site. Humans could examine samples *in situ*, onboard a pressurised rover, at a field encampment (onboard the lander or carried in a rover for temporary placement), or on Earth. Samples on Earth can be taken to state-of-the-art laboratories with equipment which cannot be taken to Mars because it is too massive and too power-hungry, or which had not been developed when the Mars mission design was frozen.[6] The final argument is that samples returned to Earth are ideal for research best carried out by several independent groups working independently – such as the Apollo lunar curatorial facility at the Johnson Space Center.

It is difficult to project ten or twenty years into the future to predict research topics and instrumentation,[6] and mission designs at present defer a decision on the hardware for scientific investigations on Mars. The assumption is that robotic exploration will change our view of Mars beyond all recognition, so there is no point planning the science for a mission based on our current knowledge. The technology developed for science on Earth will also produce changes in, for example, field science, and the technology developed for robotic missions (such as the innovative British spacecraft Beagle 2) will play a part. Technology developed rapidly as a result of the Apollo programme to land a man on the Moon 'by the end of the decade', and the same would no doubt be true for a Mars project.

All of the present studies are based on assumptions and presumptions.[10] At the moment we are accruing a large amount of data from robotic spacecraft at Mars, but even those responsible for the camera on Mars Global Surveyor – Mike Malin and Ken Edgett – have stated: 'We are constantly aggravated by the fact that all the questions we have about Mars could now be answered if we could just walk around on the planet for a few days.'[11]

Once humans are on the surface of Mars, the potential for science will be both wide ranging and intensive.

### Surface science and applications

Scientific activities on Mars will be split equally between basic science and applications.[12] The basic science would find expression in comparative planetology – comparing Earth and Mars in order to understand each a little better in terms of

During the traverses on the surface, a gradual expansion of activities will follow the early EVAs. (Courtesy Pat Rawlings.)

their present state and their history, including the origin and evolution of the Solar System. Mars and Earth formed in different parts of the solar nebula. Mars should retain clues about its formation, and its present-day surface, subsurface and atmosphere will provide clues about how and why it is different from Earth. Principal questions in basic science revolve around how the climate on Mars has changed over geological time, and the existence of past life. Climate change is as much a relevant topic for Mars as it is for Earth. Was there ever a time when conditions were suitable for life to gain a toehold? How did the earlier climate differ from what we see today? The applied science would make use of natural resources in order to sustain bases on Mars – initially for explorers, but eventually for colonists. Science will be conducted as much for operational reasons – concerning physical, chemical and biological hazards – as for pure knowledge.[13]

One of the scientific undertakings might be in prospecting for water, which could be used for manufacturing rocket propellants, or purified for use as drinking water. It has for some time been believed that water once existed on Mars, and Mars

Odyssey has discovered considerable amounts of hydrogen – probably in water molecules – below the surface of the regolith, even at the equator.

By the 1970s a new task appeared in US Mars mission planning: In Situ Propellant Production (ISPP). The wasteful process of carrying the propellants all the way to Mars, to use them only when departing the planet, could be obviated by 'living off the land' – which would also fulfil the necessity of supplying long-life base operations. This was the manner of exploration on Earth; but on Mars there is no local game to provide food and clothing, and no wood for buildings and fuel. In Situ Propellant Production was popularised by Robert Zubrin,[14] but early proponents included Konstantin Tsiolkovsky (in Russia) and Robert Ash (at the Jet Propulsion Laboratory).

*Diversity* Mars' surface features reveal a diverse, complex and long history (as extensive and diverse as Earth's), including possible volcanic eruptions through ice, with the addition of unique processes. There may also have been tectonic activity and bombardment by material from space, and there has been weathering and erosion by forces including wind, ice, and possibly liquid water. During the Apollo lunar exploration programme, a few discrete sites were sampled; but a repetition on Mars would not be viable, due to the planet's complex geological history.[5]

*Samples* The purpose of collecting samples on Mars before humans are sent there is to permit study of places where people will not be sent initially – such as the polar regions – and for engineering and safety reasons, including lander design, surface mobility, and habitability.[15]

The Apollo lunar samples ultimately reached the Lunar Receiving Laboratory on Earth, and were then sent to groups worldwide for analysis – and results are still being published. It is quite possible, however, that on Mars this work will be carried out on the planet,[6] by Principal Investigators (who receive funding for research) working as scientist-astronauts – albeit with much simpler equipment than that available on Earth. Papers might even be published in scientific journals during the return journey from Mars.

The advantage of sample analysis *in situ*, on a field traverse or at a temporary base (as basic as the landing site), is in deciding which samples are most suitable and important for return to Earth for more detailed analysis. Earth is the best location for the analysis of samples by as many people as possible with more sophisticated equipment; but there are constraints on the quantity of samples that can be carried on a spacecraft, and it will be a considerable time before the samples become available.

Due to the use of more basic equipment on Mars, the possibility of malfunctions and breakdowns will be reduced. The storage of samples could also be relatively simple. 'The best storage system for a rock is a rock' is an adage from the Lunar Receiving Laboratory.[6] It would also be easier to carry out analysis in Mars orbit, or during the journey back to Earth, rather than load equipment onto the lander, although some equipment will probably be taken to the surface. Today, much more portable equipment is being developed for use in the field or in a nearby laboratory, including electron microscopes and equipment for analysing the composition of rocks. On Mars, this could include a search for microfossils – which humans could carry out much more efficiently than robots.

*Biology* The primary purpose is to search for life which might exist on Mars today.[16] The first plans for spaceprobes and human missions assumed that plant life existed on the surface, because Mars' changing features resembled the growth and demise of large areas of vegetation. If there is life there now, it will be microbes. These are the simplest forms of life, and it is almost certain that there are no life-forms on Mars larger than viruses or bacteria. After the Viking landers found no trace of extant life in Mars' hostile environment, the search switched from present-day life on the surface to life far underground or in caves (as is now found on Earth in the most extreme environments), and also extends to fossils of ancient microbial life in the rocks. Any fossils will be microfossils or even nanofossils.

*Palaeontology* encompasses geology and biology. It is difficult to find ancient signs of life on Earth, and it will be an even greater challenge on Mars. The oldest parts of Earth's crust can be found in Greenland and Australia in particular; but a large part of the surface of Mars (excluding volcanic regions such as Tharsis) is ancient. Old rocks are therefore much more abundant, but it will still be very difficult to find microfossils. The real challenge will be to find irrefutable evidence. Supposed indications of 'bacteria fossils' in the martian meteorite found in the Antarctic (ALH84001) remain controversial, even though the claim sparked the apparent discovery of live bacteria in crustal rocks on Earth. On Mars there will also be a search for biomarkers – an indication that life exists somewhere today, or that it existed in the past.

**The geosciences**
Geoscience falls into three main categories (although in strict terms the prefix 'geo' refers to the Earth, whereas the equivalent terms relating to Mars should be prefixed with 'areo'.)

*Geology* Rocks from the crust are easy to find, but mantle rocks are more elusive, although it should be possible to find them in locations where the subsurface has been exposed, such as impact basins or craters. Samples of deep crustal and mantle material may be found at locations of meteoritic impact, which have produced ejecta. The other option is deep drilling. Just as meteorites provide clues about the material in the planetesimals that were the building blocks of planets, the material excavated by large impacts is just as important for understanding the interior of a world. They are volcanic igneous rocks, sedimentary rocks laid down by wind or water, and metamorphic rocks (igneous or sedimentary rocks that have been changed by heat and pressure). As on the Moon, explorers might find new minerals.

*Geophysics* maximises the use of instruments deployed by humans. The measuring and mapping of areas of heat flow from the planet's interior can provide clues about its structure. After drilling into the surface, heat-flow probes will be emplaced to measure the amount of heat released by the interior of Mars to reveal variations in the thickness of the lithosphere at different locations. Seismological studies will provide further clues as to the structure of the planet's interior. Magnetometry will involve measurements of Mars' magnetic field (or rather, its 'fossil magnetic field'), at a higher resolution than is possible from orbit. Gravimetry (mapping of the gravitational field) will produce data on the structure of the subsurface, which also will be of higher resolution than measurements from orbit. Electromagnetic soundings will utilise ground-penetrating radar.

*Geochemistry* Geochemical analysis can be carried out (to a limited degree) by instruments on robots – orbiters, landers or rovers – or in a field laboratory. However, some reports have emphasised the importance of analysing samples in laboratories on Earth, to take full advantage of state-of-the-art equipment.[15] – which is particularly important in the dating of rocks.

### Weather and climate
Meteorology is the study of current weather, while climatology is concerned with long-term trends. Mars has a thin atmosphere, but it is very different from Earth's atmosphere. There are no oceans to complicate atmospheric cycles, but a significant part of the atmosphere condenses at the pole in the winter – a phenomenon quite unlike any on Earth.

Meteorology will form part of the work on the surface for operational reasons, but the long-term climate will also be studied. A weather station would typically take

Working on the surface of Mars during remote EVAs may include activities during a mild dust-storm. (Courtesy Pat Rawlings.)

measurements of temperature, pressure, moisture content, wind speed and direction, and clouds. The past climate can be studied by drilling out deep core samples, in the same way that ice-cores on Earth reveal the state of the atmosphere when a particular part of the ice became frozen.

Mars is believed to have once had a much thicker atmosphere, higher surface temperatures, and stable bodies of liquid water. The resultant weather system would have led to erosion of the surface, the formation of ice (causing more erosion), and the deposition of layers of sedimentary rocks. As Mars has no large moon, its axis 'wobbles' over comparatively short periods, and this would lead to major climatic shifts. (Earth's axis is also subject to nutation, but over longer periods, as it is regulated by the Moon.)

### Upper atmospheric physics
On Earth, the upper atmosphere consists of the troposphere and the stratosphere, and solar particles are deflected around the magnetic field. Mars also has a lower, middle and upper atmosphere,[17] but its magnetic field is so insignificant that the solar wind effectively scours the atmosphere. Mercury is a much smaller world, but has a magnetic field; whereas Venus has no intrinsic magnetic field, but rather an induced field in the ionosphere, caused by the solar wind. It is not clear how a planetary magnetic field is created, but the accepted theory is that it is generated by the 'churning' of conductive material in the core. Aurorae are caused by the solar wind acting in the upper atmosphere, but it is not known whether they occur on Mars.

### Dust
Dust can cause equipment failure by clogging components, and airborne dust and the regolith poses a mobility problem for people and vehicles. Filters become clogged, helmet visors are scratched, spacesuit seals deteriorate, vehicles can sink

Extended traverses may also include expeditions to prominent surface features. (Courtesy Pat Rawlings.)

into the surface material, and tyres wear out. Dust can cause electrostatic effects and affect communications, it reduces visibility, it increases wear and tear of equipment, and it can also cause medical problems. Solar panels are effective on Mars, despite the weaker sunlight, but they are hindered by dust.

## Radiation

Because Mars has only a thin atmosphere it suffers higher exposure to cosmic and solar radiation, and its negligible magnetic field does not deflect charged particles. A radiation monitor is therefore of prime importance on the surface. It was to be sent in 2001, but the lander was cancelled after the debacle of the Mars Polar Lander mission in 1999. Radiation levels near Mars were found to be 2.5 times those at the ISS in Earth orbit. After increased solar activity, the level of radiation at Mars also increases dramatically. (As an example, Mars Odyssey carried an instrument that plotted radiation levels all the way from Earth to Mars and in Mars orbit. During autumn 2003 the Sun was so active that it destroyed the spacecraft's radiation monitor.) Humans on Mars will need to 'duck and cover' as quickly as possible.

## Equipment

On a mission to Mars – especially for a 1–2-year stay on the surface – spare parts need to be carried. Some of today's equipment is very complicated, but something simpler is required – which is why field geology is the ideal science discipline for Mars. All that is required is a human with a hammer. If the hammer breaks, a makeshift hammer can be made from a length of pipe.[10]

The other advantage of human missions is that vehicles can be used.[18] Science can be carried out on the move, using a pressurised rover carrying simple equipment. The only disadvantage of a moving vehicle is the possibility of damage and necessary repairs. On Mars, there could be an open-topped unpressurised two-seater rover (like the Apollo LRV) and a pressurised laboratory rover for field-trips. The latter would be a mobile version of the lander habitat, capable of covering distances of tens of kilometres, possibly up to 100 km. With humans on Mars, telerobotics is feasible. The robots would be left to fend for themselves, but in case of a delicate or complicated task, humans would control them from a distance, using virtual reality systems. They could be drilling cores, scraping rocks and collecting samples thousands of kilometres from the landing site or from the nearest human. Such rovers could be sent in advance to stockpile samples or to take samples to the lander laboratory after humans arrive on the surface.

## SURFACE SCIENCE PACKAGES

On Mars, astronauts and robots would deploy instruments for long-term areophysical and meteorological measurements – a follow-on to the Apollo Lunar Surface Experiment Package (ALSEP).[9] The first Apollo landing omitted any complex science packages – not only to avoid overburdening the astronauts, but because the landing itself was the most important (engineering) aspect of the mission. This led to the creation of the Early Apollo Surface Experiment Package (EASEP), based on the full ALSEP science package that was taken on later missions. The journey to Mars is such a challenge, however, that as much as possible would be carried out on the first landing – and it might be the *only* human mission to Mars for some time. Simple science packages might be included, but there could also be expanded science packages ready in Mars orbit for later landings on the same mission, assuming that there would be more than one landing during the mission. Another possibility is that robotically deployed packages would be sent first, and that a long stay, as part of the first human landing, would allow the deployment of more complex packages. Teleoperated robots could emplace science packages without human presence, with the human operators situated elsewhere on or near Mars, and the data would be sent to the mother-craft in Mars orbit or direct to Earth. The advantage of having humans on the surface, however, is that they can position and level the packages, which might be carried in a trailer attached to the pressurised rover.

The Mars Science Experiment Package (MSEP) is primarily concerned with meteorology and geophysics, including heat flow, seismology, magnetometry, gravimetry, and subsurface radar.

### Comparison of ALSEP and MSEP

| ALSEP Number | Description | Discipline | Apollo | MSEP Mars |
|---|---|---|---|---|
| S-031 | Passive Seismic | Geophysics | 11–16 | Yes |
| S-033 | Active Seismic | Geophysics | 14, 16 | Yes |
| S-034 | Lunar Surface Magnetometer | Geophysics | 12, 15, 16 | Yes |
| S-035 | Solar Wind Spectrometer | Geophysics | 12, 15 | Yes |
| S-036 | Suprathermal Ion Detector | Geophysics | 12, 14, 15 | Yes |
| S-037 | Heat Flow | Geophysics | 13, 15–17 | Yes |
| S-038 | Charged Particle Lunar Environment | Geophysics | 13, 14 | Yes |
| S-058 | Cold Cathode Ion Gauge | Geophysics | 12, 14, 15 | Yes |
| S-059 | Lunar Field Geology | Geology | 11–17 | Yes |
| S-078 | Laser Ranging Retro Reflector | Astronomy | 11, 14, 15 | Yes |
| S-080 | Solar Wind Composition | Astronomy | 11–16 | Yes |
| S-151 | Cosmic Ray Detection (Helmets) | Astronomy | 11 | Yes |
| S-152 | Cosmic Ray Detection (Sheets) | Astronomy | 16 | Yes |
| S-184 | Lunar Surface Closeup Photography | Geology | 12, 13 | No |
| S-198 | Portable Magnetometer | Geophysics | 14, 16 | Yes |
| S-199 | Lunar Gravity Traverse | Geophysics | 17 | Yes |
| S-200 | Soil Mechanics | Technology | 14–17 | Yes |
| S-201 | Far UV Camera/Spectroscope | Astronomy | 16 | No |
| S-202 | Lunar Ejecta and Meteorites | Astronomy | 17 | No |
| S-203 | Lunar Seismic Profiling | Geophysics | 17 | Yes |
| S-204 | Surface Electrical Properties | Geophysics | 17 | Yes |
| S-205 | Lunar Atmospheric Composition | Geophysics | 17 | Yes |
| S-207 | Lunar Surface Gravimeter | Geophysics | 17 | Yes |
| M-515 | Lunar Dust Detector | Technology | 12–15 | Yes |
| S-229 | Lunar Neutron Probe | Geophysics | 17 | Yes |

Many of the experiments carried in the lunar science packages would have equivalents on Mars missions, while others would either be heavily modified, possibly replaced by something entirely different, or simply not be included:

*Surface electrical properties* This instrument used the equivalent of ground-penetrating radar to sense what was below the direct surface layer of the Moon by looking at the properties of materials there. This would be equally applicable to Mars missions.

*Laser ranging retro-reflector* This will not be used on Mars, due to the Earth–Mars distance and the atmospheres of the two planets. There is an advantage in detailed study of the orbits of Phobos and Deimos, and lasers could be placed on them and aimed at a receiver on Mars, or vice versa.

*Cosmic-ray detection* Cosmic rays reach the surface of Mars, but the main reason for such research is operational; for example, as part of total radiation exposure measurements.

*Ultraviolet spectroscopy* will not be relevant on Mars, as the atmosphere, although thin, will block much of the ultraviolet radiation from space. The main subjects of study will be Phobos and Deimos.

*Solar wind spectrometer* Much of the solar wind at Mars is blocked by the thin atmosphere, but its effect on the atmosphere and on the surface could be studied.[19]

*Meteorology* Mars has weather, and so meteorology will be important – chiefly for operational reasons, but also for studies of the long-term climate. There are no oceans on Mars, and the weather systems are (in theory) less complex than those of Earth; but 20% of the atmosphere changes from gas to ice at the poles each winter!

*Heat flow* is measured with platinum resistance thermometers placed inside holes drilled several metres into the ground. Heat is exuded from the possibly still molten core of the planet and from the decay of radioactive elements in subsurface rocks. To avoid interference with the readings, the holes must be bored sufficiently deep so that the thermometers are isolated from surface heat. The area around the instruments can take several months to completely dissipate the heat caused by the drilling. Heat flow is a fundamental property of any world, and detectors are deployed to form a local (closely spaced) or regional (widely spaced) picture.

*Seismometry* Active and passive seismometers are used in deep and shallow studies of the planet's interior by sensing movement in three axes. The sound waves passing through the rocks are termed surface waves, S waves and P waves. S waves are analogous to a 'slinky spring' moving from side to side, while P waves are resemble a push/pull action. Such motions are due to internal and external forces. Internal forces on Earth, for example, arise from plate tectonics, and external forces include the impact of meteorites and tidal forces caused by an orbiting moon. This is passive seismometry, in which detectors are placed on thermally stable ground. Active seismometry uses a deployed network of detectors, in conjunction with explosives detonated to produce sound waves which are measured after being reflected off subsurface layers. On Earth, geophone networks are used in prospecting for oil and gas, and seismometry was incorporated in the Apollo lunar missions.[20]

*Gravity* Where there is mass, there is gravity. According to Einstein's theory, gravity is a distortion of spacetime caused by mass; but whatever the interpretation, more mass results in more gravity. The Earth has more gravity than the Moon, and Mars' gravity is somewhere between the two, but the gravitational field is not uniform. Gravity varies on different parts of a planet, and the variations can be measured with a gravimeter, in milligals (differences from the average in thousandths of a Gal, with a Gal being 1 cm per second squared. Scandinavia is a lower-gravity area, because the land is slowly rising after being crushed by ice sheets during the last ice age. On the Moon, mascons (mass concentrations) – which are predominantly situated in the lunar maria – are associated with higher gravity, caused by either lunar mantle material that has risen, or relatively massive lumps of basalt. Apollo 17 carried a Lunar Traverse Gravimeter to study the subsurface in the Taurus–Littrow valley. The instrument was carried on the Lunar Roving Vehicle, and was deployed on smooth ground at several points during the EVAs. The astronauts had to remain relatively motionless so as not to disturb its 3-minute-long data-gathering runs; but maximum sensitivity was not achieved, due to an instrument malfunction. Gravity data is notoriously difficult to interpret.[21]

*Magnetic fields* Planetary magnetic fields are believed to be caused by convection currents near the molten metallic core. They are bipolar fields, and they change with time. For example, Earth's poles are slowly moving as measured at the surface, and the field is becoming weaker. Eventually there may be no magnetic field at all, and when it returns the magnetic poles will have been transposed. The magnetic field in a given location is preserved in rocks that were once molten, and can be measured with a magnetometer. The discovery of these 'fossil' magnetic fields led to the theory of plate tectonics on Earth, by which the continents are slowly moving around the crust.[20]

*Soil mechanics* includes studies of, for example, unstable ground that will not support the weight of a human (the so-called martian 'foo-foo dust'), and practical experiments, such as rolling rocks into craters to study the resultant trails.

*Water* Water molecules in the air at the surface would indicate subsurface deposits of water. Ground-penetrating radar will also sense deposits of subsurface water-ice. The sensors on satellites such as Mars Odyssey have detected hydrogen in the martian regolith, and this may be in the form of water.

*Drilling* As with Apollo there would be drilling to shallow depth, but occasionally there would be an attempt to drill much deeper, to locate and tap into reservoirs of ice or water, or to investigate the stratigraphy of the area. Drills can penetrate below the surface oxidisation layer caused by ultraviolet light and oxidising chemicals, to approach any water table which might exist. The topmost layer on Mars is very oxidising, and it is where organic chemicals are destroyed. Mars should be covered with organic chemicals from comets, but they appear to have been destroyed, and the surface is also completely barren of microbes. Drilling is another justification for sending people to Mars. It is best performed by humans, the equipment is large and power-hungry, and it can break down. Robots may start the drilling after assembly of the equipment by humans, but when a problem arises, the astronauts would become involved. This is one example of the teamwork between robots and humans. Drilling would be very slow in order to avoid damage due to heating, and lubricants would have to be avoided, as they might harbour contaminants from Earth. The same problem exists with deep drilling in the Antarctic, where the large subsurface Lake Vostok has been cut off from the outside world for 35 million years.[22]

*Cores* Obtaining sections of the soil in a given location allows analysis of its layers to provide information on the geological history of the area. On Apollo 11, core samples were obtained simply by pushing (and hammering) a hollow tool less than a metre into the ground. At the other extreme, a power drill-head was used on Apollo 15–17. A hollow rotary-drill core allowed collection of samples from a depth of up to 3 metres, but on Apollo 15, Dave Scott experienced problems in trying to pull the core out of the ground. The bore stems were different from those used for the heat-flow experiment, but they caused similar problems. They could be sunk to their planned depth of 2.4 metres, but the drill flutes became clogged with cuttings. To complete the task, Scott was assisted by Jim Irwin, but he overworked his life support system and strained his back and hands. For Apollo 16 and 17 the apparatus was modified, and the treadle on which the astronaut stood to support the drill was fitted with a jack to lever the drill-stems out of the ground.

However, the troubles associated with the drill stems continued. A 'vise' on the rear of the lunar rover was supposed to hold the core tubes so that they could be split into sections for transport back to Earth; but unfortunately it was fitted backwards and would not grip the core sections, which had to be held by hand. The deep core was regarded as so valuable that exploration of a possible volcanic area called North Complex was eschewed in order to remove the deep core from the ground and split it into sections. Later analysis of the core revealed more than fifty distinct layers – a record of the multiple events that had created the column of lunar soil.

The method of operation was to assemble two 0.5-metre long bore stems, with a simple push-joint between them. They were then inserted into a drill chuck, and the battery-powered drill was used to drill them into the surface. When about a third of their length was left above the surface, a wrench was used to release the chuck, and the drill was removed. Two more bore stems were then joined together and pushed into the pair protruding from the ground, and the drill chuck was reset and the drill fixed atop them. Drilling then resumed, and the entire exercise was repeated for a third pair of bore stems.

Apollo's 3-metre deep cores far surpassed anything obtained by robotic drillers. The Venera probes on Venus drilled to only a few centimetres – but in extreme conditions of temperature, pressure and chemistry. The Soviet Luna probes did not fare much better. The first two (Luna 16 and 20) carried a deployable 90-cm drilling rig that managed to collect cores to depths of 35 cm and 15 cm respectively. However, Luna 16 hit rock, and drilling was stopped to prevent damage, Luna 20 also struck rock, and the drill motor cut out automatically. A stereo camera on a tall boom took a photograph of the locale to place the core sample in context of local geology; but the first landing took place during the lunar night, and as there is no evidence that floodlights were carried, it is improbable that photographs were obtained. The drill could be swept around up to 100° by electric motors commanded from Earth, and the drill head could swivel; but the arm had a fixed length, and was far less flexible than Viking's extendable sample-collection scoop. On the last two missions, Luna carried a combined percussion and rotary drill that would hammer into soft soil and switch to rotation if it met resistance. The combined drill rig was fixed to the side of the lander, and could drill to a depth of 2 metres. Luna 23 landed badly on the Moon, but Luna 24's drill obtained a 1.6-metre core sample. It was pushed up into an 8-mm-diameter flexible tube that coiled like a spring.[23]

### Lander science laboratory

The main reason for having a laboratory facility on the Mars landing craft would be for the alleviation or prevention of forward and back contamination. This is often referred to as 'planetary protection' – although for the general public the planet is normally Earth. The Biological Isolation Garments (BIG) worn by Apollo astronauts when they returned to Earth highlighted the issue of contamination. The danger of succumbing to infection from the martian air, water or soil (back contamination) is real, even if improbable; but the astronauts might equally take viruses, bacteria or biochemicals (even oxygen) from Earth to Mars. If a bacterium were to be found on Mars, would it be indigenous, or from Earth?

### Laboratory design

The design of the lander laboratory would depend crucially on the first scientific results from the robotic missions. It might develop into being like the Lunar

A 1980s interpretation of expanding the landing site to include additional modules for science or habitation. (Courtesy Pat Rawlings.)

Receiving Laboratory – a very complicated and almost hospital-like facility. This, in turn, would affect the design of the landing craft. It will add too much mass and complexity to the main landing craft, so it might be a facility launched and landed separately and then attached when on Mars. The Russian Mir programme incorporated studies of a Medilab module, and NASA studied the Antaeus quarantine facility for martian sample analysis in Earth orbit.

### Sample analysis in the laboratory

Many samples will be collected by robots, which may take them to the landing site even before the landing craft has touched down. The results would be used in determining the best sites for human landings, and the location of areas of particular interest.[9] Throughout the course of the mission, samples of rocks and regolith would be collected during EVAs and returned to the base camp or to a large pressurised rover. Robots under the control of humans could also collect samples from many parts of the surface. Unpressurised rovers would probably be used for short-range EVAs, and large pressurised rovers would be used for local or regional sorties. The surface samples and atmospheric samples would then be analysed to determine mineral content, chemistry and geochronology, and the results would be used in the planning or reassessment of EVAs or sample retrieval by robots. Information will be transmitted to Earth (probably daily), for assistance in replanning.

### Sample collection on Mars

Teleoperated rovers could collect samples from thousands of kilometres away, and humans could transport many kilogrammes of samples to a laboratory at the lander,

A longer stay could include further expansion to incorporate additional pressurised or unpressurised facilities. (Courtesy Pat Rawlings.)

where they could be studied in a sterile mini-laboratory. A field geologist would then decide which of them should be packaged for return to Earth.

Astronauts working with glove-boxes and hands protected by gloves in the field laboratory will avoid exposure to any possible microbes on the samples, and the long journey home would provide time enough for any reaction to the samples to become apparent. Procedures developed by NASA to comply with planetary protection rulings from 1968 (as part of a United Nations treaty) were used on the Apollo missions. Samples are exposed to biological materials and to animals such as rats, to determine whether there is any adverse reaction. The lunar samples smelled like gunpowder when the sample boxes were opened in the lunar module, and martian samples will have to be tested in advance for toxicity if there is a risk of inhalation by the crew within the landing module. There will also need to be tests to ensure that there is no danger to the integrity of the landing craft.[6]

## Microscopes

Microscopes are chiefy used in the biological sciences, but are also used in geological research. Microfossils, and perhaps even algae and lichens, can be sought in slivers of Mars rock, but life as we know it is unlikely to exist on the surface of Mars today, because of the high levels of radiation and oxidising chemicals.[6]

The electron microscope works by firing a parallel beam of electrons at the object to be studied, and the interactions are detected and recorded on a cathode-ray tube. This type of miscroscope is capable of revealing objects at nanometre resolution. The scanning electron microscope uses a finely focused beam of electrons scanned across a bulk sample, and the intensity variations form a three-dimensional image of the surface. The transmission electron microscope operates in the same way as an ordinary slide projector. The light source is a focused beam of electrons, and the image is recorded on a phosphor screen.

## Learning and returning

Samples gathered on the martian surface will be stored in zip-lock bags and packed in containers, (along with samples of the atmosphere) ready for shipment to Earth. The advantage of having analytical tools *in situ* is that it can be decided what is worth bringing back, as the quantity will be limited by total lift-off mass and mass distribution for re-entry.[6]

There will be problems if dust-storms or dust-devils roll through an MSEP site, even if they affect only the high-gain antenna. After the departure of humans, repairs will be possible only if robots are left as caretakers.

Outreach during a long stay on Mars is a challenge, to convince people that pure science is interesting. It costs money for people to work in, for example, Antarctica throughout the entire year, and the scientific work carried out there generates information on subjects which could directly affect the general public (such as ozone depletion). Antarctic bases were originally set up for political and strategic reasons. The first British base was established in the 1940s, when it was feared that German commerce raiders would be based there, and science was a secondary consideration.

The Apollo programme followed the same reasoning, but we still need firm justification for a journey to Mars.

## SCIENCE ON TRAVERSES

There will probably be at least one pressurised rover for use far from the landing site, and even a 1-month sojourn on Mars should be able to accommodate at least one week-long traverse. There will also be small unpressurised rovers for use in close proximity to the landing craft science laboratory. The last three Apollo missions each carried a Lunar Roving Vehicle so that science could be carried out across a region and not just in the immediate vicinity of the landing site.

Four options are available for scientific work during or following a traverse: at the lander science laboratory, using samples gathered on the traverses; during field traverses by geologists; scientific results carried and left on the rover during a traverse; and science packages (MSEPs) carried by the rover, and emplaced in a location. The rovers would have their own power source, and would communicate during the stay or for many years after the humans have returned to Earth. Instrument packages would be placed by astronauts and robots for long-term geophysical and meteorological measurements – a follow-on to the Apollo ALSEP packages.

### Size of the field party

There is no ideal size for a field party, but two or more is best for safety, although they would probably be working independently. They may well leave one person in the landing craft or science laboratory to carry out analysis for feeding into the next day's traverse, and one in the landing craft or habitat for communications or IV mission control duties. Almost nothing except safety is gained by having more than one person examining the same object in the field. The only other benefit would be if assistance was required. On space station EVAs, the EVA team works as a pair, carrying out independent tasks, but watching each other's backs and maintaining constant communication. Traverses will not be attempted every day, and the crew will perhaps alternate in pairs between sample analysis and traverses. Anyone injured during the course of a traverse would need to recuperate in the habitat, landing craft or science laboratory. However, communication with those outside would still be possible, and the 'patient' might even be able to undertake IV or communications duties. This interaction would be recorded, and ultimately transmitted to Earth.

### Other considerations

*Medevac during traverses* In the event of an accident it would be difficult to examine someone who is wearing a pressure suit, and they would have to be taken to a pressurised facility such as the lander. A rover would probably carry a variant of the Shuttle Orbiter Medical System (SOMS), and the landing craft or habitation module would carry a variant of the ISS Crew HealthCare System (CHeCS).[24]

*Lighting* Mars has a normal day/night cycle of around 24½ hours. Some of the daylight will be lost according to the season, and working at night under arclights and spotlights may be part of the mission or traverse specifications. The suits, however, will probably be equipped with lights (as was the Shuttle and ISS EMU spacesuit).

*Unique rocks* We have no 'hands-on' experience of rocks created and changed over millions or billions of years in an environment subjected to high levels of radiation and with low levels of oxygen, high levels of carbon dioxide, and low gravity. Even gravity alone might change the terrain. Piled material slips and slides under conditions different from those experienced on Earth. All elements are universal, but the minerals on Mars may be different, and might have been affected by the presence of life, which changes the chemistry and leaves deposits.

*Artifacts* As rovers become more advanced, new models will be sent to Mars; but the old rovers will remain there – and might eventually become exhibits in a museum on Mars. The landers on Mars include the Thomas Mutch Memorial Station (Viking 1 lander), the Sagan Memorial Station (Mars Pathfinder), and the Columbia and Challenger Memorial Stations (the MER landers).

*Tools* Tools will be as simple as possible. For geology, only a hammer is required – and if it breaks, a piece of metal or a rock can be used. Conversely, sophisticated portable sensors, available for use on Earth, might be used on Mars. (In forensic science, for example, field analysis is beginning to predominate over laboratory analysis.) However, on Mars it will not be particularly easy to bend to pick up samples (as when wearing the early lunar EMUs), and so relatively simple tongs and extensions will still be required.

*Ground truth* One very useful aspect of a Marswalk is ground truth: a comparison of ground-level observations and observations from the air or from orbit – either on the mother-craft or by satellites. The satellites might be primarily occupied with monitoring the weather or mapping resources, but the combination of the view from above and the view from the ground is beneficial. Samples from a particular location can be compared with instrument readings from the air or from space and this can then be applied to similar signatures on other parts of Mars.

**Justification for human scientists on Mars**
The 1998 Mars Field Geology Workshop (see Ref. 20 under 'Evolution of a Marswalk') tabled three Big Questions:

*Big Question 1* Why explore space at all? 'Because the future and the whole universe lie out in front of us.' Human beings, by their very nature, cannot help but wonder what is 'out there', whether it be our local neighbourhood or deep space. The team suggested that it would be far better to take well-considered strides into the future, rather than be passive.

*Big Question 2* Why go to Mars? By recognising that certain processes on Mars are similar to those on Earth, but on a different scale, differences between the planets are

The benefits of a human crew for exploration could include investigations in areas that are difficult for automated vehicles to reach, although risk and safety will have to be assessed. (Courtesy Pat Rawlings.)

more fully revealed. This will help to establish further understanding of the uniqueness of Earth and, by implication, commit us to its protection. The surface area of Mars' terrain is roughly equivalent to the land-area on Earth, and the team suggested that human missions to Mars could incorporate a degree of 'living off the land', so that living on Mars could become more 'homely'.

*Big Question 3* Why send humans to Mars? The human mind is amazingly good at detecting key elements in an otherwise overwhelming accumulation of data – spotting in an instant what is out of place, or unique to a given scenario. Millions of years of evolution has provided humans with capabilities that robotic intelligence or remote sensing could not, as yet, possibly supersede. Despite pitfalls and human frailties, the experienced scientist on site will quickly produce accurate and reliable results. Such knowledge can be gained only from experience, and not programming. This is never more true than in field geology.

## REFERENCES

1 See *A Catalogue of Apollo Experiment Operations*, NASA RP-1317.
2 Benton C. Clark, *Crew Activities, Science and Hazards of Manned Missions to Mars*, Planetary Sciences Laboratory, Martin Marietta Astronautics, Denver, Colorado, 39th Congress IAF, Bangalore, India, 8–15 October 1988, IAF-88-403.
3 Carol R. Stoker, *Science Strategy for the Human Exploration of Mars*, AAS 95-493.
4 *Assessment of Mars Science and Mission Priorities*, NRC 2003.
5 Michael Carr, *Scientific Objectives of Human Exploration of Mars*, AAS 87-198.
6 Interview with Michael Duke, 28 January 2004.
7 Interviews with Bo Maxwell, 5 and 16 December 2003.
8 Interview with Charles Frankel, 9 January 2002.
9 Interview with Alex Ellery, 10 October 2003 and 4 February 2004..
10 Interview with Graham Worton, 4 December 2003.
11 Kathy Sawyer, 'Mars,' *National Geographic*, February 2001.
12 Benton C. Clark, *Manned Mars Systems Study*, AAS 87-201.
13 *Safe on Mars*, National Academy Press, 2002.
14 Robert Zubrin, Mars Direct.
15 *Planetary Exploration through Year 2000: an Augmented Program*, NASA SSEC.
16 C. Mckay, 'Scientific goals for martian expeditions' *Journal of the British Interplanetary Society*, **57**, no. 3/4.
17 Joseph M. Boyce, *The Smithsonian Book of Mars*.
18 Benton C. Clark, *Mars Rovers*, AAS 95-490; Carol R. Stoker, *Science Strategy for the Human Exploration of Mars*, AAS 95-493.
19 Gene Simmons, *On the Moon with Apollo 16*, NASA.
20 Gene Simmons, *On the Moon with Apollo 15*, NASA.
21 Gene Simmons, *On the Moon with Apollo 17*, NASA.
22 Interviews with Michael Duke, 28 January 2004; Ian Crawford, 15 January 2004; Charles Frankel, 9 January 2002.
23 Andrew Wilson, *Solar System Log*.
24 *Report of the 90-day Study on Human Exploration of the Moon and Mars*, NASA, Section 6, 'Meeting human needs in space' (referring to the Space Station Freedom Health Maintenance Facility).

# Living the dream

Even though there has been no official programme to send humans to Mars, material has been written and published on the methods and procedures for journeying there and back, as well as on some of the obstacles that would need to be overcome before we can send people. In this book we have assumed that humans will one day walk on Mars, and have concentrated on what the crew may attempt to do after landing, and the conditions that they might encounter.

However, one vitally important factor will need to be addressed in order to ensure the success of the mission, regardless of methods and objectives: the health and well-being of the crew involved. On such an extended and potentially dangerous mission, what do we need to do to help the crew to cope with the psychological, emotional, physical and possibly even cultural burdens of being so far away from home for so long, and how do we ensure that everyday basic human needs, such as food and hygiene, can be catered for during the entire mission?

## CREW TRAINING

It will take many years to prepare for a mission to Mars. On Apollo, generic training took 1–2 years – and that was after basic training, general training, survival training, academic training, flight training, and general systems training. Training for a specific Apollo mission depended on how the astronaut progressed through the mission cycle. If he first became a member of the support crew, and than acted as back-up before moving on to a prime crew position, the mission training could be shortened, as he would have already passed through the training system. The Russian system of crew training is to select a group of cosmonauts for a specific flight, train them as a group, and then select the flight crew from among them shortly before launch. In this way it is easier to replace crew-members in the event of illness, as there is a core group of experience to draw upon.

On the Shuttle, the training time was stretched a little to include Ascan (Astronaut Candidate) training and then advanced training, followed by mission-specific training tailored to the astronaut's role on the crew. On Mir space station missions with the Russians, the Americans also had to learn the language. On the ISS the common language is English, but some of the astronauts can speak Russian, Japanese, French, German and Italian. The same will apply during the journey to Mars – an international crew with a common language (probably English) and skills in each other's languages. A space station mission normally takes 2–3 years for training in different phases, and a Mars mission will certainly require a long-term training programme (possibly 5–10 years) because of the amount of information needed. The crew will have to operate and maintain the vehicle while journeying to the planet, and deal with orbital operations, rendezvous and docking, observations, orbital rescue and contingency training, repair and maintenance, habitability issues, and then, on the surface, the EVA systems, the lander systems, the science equipment, the control of rovers and the teleoperation of robotics.

Most cosmonauts consider that no-one can be called a true space explorer until he/she has orbited the Earth, and that the best place to train for spaceflight is in space itself. Therefore, training might also include a term on the ISS to become accustomed to a weightless environment (sickness, crew psychology) and possibly a mission to the Moon, but training must not be excessive. Mission delays have previously caused problems in maintaining peak performance, and a Mars crew will not remain in peak condition, constantly, for 2–3 years. They will encounter periods of low morale, emotional peaks and troughs, and other problems – and some of what they have trained for will not even take place until several months into the mission.

This is a case for in-flight training in certain aspects of the mission, including computer program upgrades or equipment modifications, plus simulations and virtual-reality training and refresher courses. Spaceflight history already demonstrates the importance of these procedures: for example, the collision of Mir with the Progress freighter in 1997. Mir-23 Commander Vasily V. Tsibliyev had to attempt to dock the Progress using the TORU system, but he was unfamiler with the procedure and had not practised it for some considerable time.

Crew compatibility and working as a team will be vital for the mission, even before launch, because some crews which trained as a team had to cope with injury, ill health, death, retirement or reassignment during the training period, and the problem of accustomising themselves, and working with, a replacement crew-member. During the long period of training that will be necessary for a Mars mission, this is even more likely to happen.

The style of training must also be considered. A mission to Mars will incorporate a degree of autonomy purely because of the distances and time delays involved, and with a possibly international crew these circumstances favour the Russian approach to training over the NASA approach. Throughout NASA's programmes, training has usually consisted of constant rehearsal and strict timelines, whereas the Russians have tended to train people to live in space and then left them to determine the best way of doing so. On the ISS, this rigidity (of US crew-members) has relaxed somewhat because of the length of the missions. The same will apply for a Mars mission.

Apollo EVA training was timed to the minute. The astronauts had a map of the surface, and had to move from place to place within a certain period of time. They had to deploy a particular experiment at a certain moment in the EVA, and they had to move on if they ran out of time. On Mars it would be very difficult to do this, and practical field geology would have to be different. There will be no direct communications with Earth, and the astronauts will not be able to tie down every minute of every surface activity, because they will not be able to detail every EVA beforehand. They will be reacting to what they have found. The first EVA will certainly be time-constrained to ensure that the systems work, but subsequent EVAs will rely on accrued experience. Training for every minute of every EVA will be impractical, and probably will not happen.

Earth-based training in simulators and remote terrain mock-ups will in some cases be supplemented by experiments in undersea habitats and familiarisation with long-duration submarine operations, as has been done by several of the current ISS crew-members. There will certainly be training sessions in remote locations such as those currently operated on Devon Island (in Canada) or Antarctica. After being assigned to a 'Mars crew', some 'training' flights onboard the ISS (or its replacement), and possibly to the Moon, will help the astronaut to become familiar with working patterns and acclimatisation procedures for enclosed environments, zero or reduced-gravity operations, and crew compatibility, to develop the 'team' approach and 'hands-on' experience in spaceflight environments and operations. These spaceflight training missions will help in accruing habitability experience that will be valuable on the early stages of the flight to Mars.

Physical fitness will also be a vital aspect of training. The crew will not necessarily have to be able to complete a triathlon, but the fitter they are, the better they will be able to withstand launch and recover from long periods in microgravity. Exercise is also linked to their diet and their psychological state, and it will be very important to maintain not only their training and peak performance in operating the machines, but also their bodies over a long period of time. Exercise will continue during the flight to Mars, and will be essential in recovering sufficiently quickly to be able to operate on the surface as soon as possible. Astronauts use a variety of training methods – such as the treadmill and running track, rowing machines, and other types of equipment – but if this has to be included in the lander, for example, then it will add to the landing mass and reduce the amount of space available for other equipment. A gymnasium taken to Mars will probably be small and compact.

Weight training will probably also be included, because once the travellers reach the surface of Mars they will be lifting heavier weights, and they will want to maintain their physical condition in order to operate efficiently in pressure suits. In addition, during a year's stay on the surface this will help recondition their bodies in preparation for the long flight home. If there is a weight system gym inside the vehicle, they could, on a day off, spend an hour or two in physical training, have a shower, relax, and maintain physical condition, linked to calorie requirements and dietary needs.

As stated above, the best experience of living and working in space is actually being there, and the long flight to Mars will allow plenty of time to develop an effective and efficient work programme that can be adapted for the return to Earth. However, a new learning curve will be required soon after arriving on the surface of

Mars, and will probably be developed during acclimatisation to the gravity environment prior to the commencement of surface operations. Equally, there will have to be an adjustment period prior to embarking on the return journey.

**Hardware**

The facilities on the mother-craft and lander will have to await defined programme decisions, but experience gained on Russian space stations, Skylab and the ISS has provided some data on the operational use of spaceflight hardware over many years in space. However, this has always been in low Earth orbit, and man-rated hardware has yet to be tested over the distance to Mars. The only experience we have of human exploration in deep space is the 'short hop' to the Moon,

Presumably, some precursor missions carrying Mars hardware will be evaluated prior to dispatching humans that far into space. Boilerplate and unmanned versions of manned craft have been used in several programmes, but the Shuttle completed its first space mission with a crew onboard. Despite years of research into the aerodynamic characteristics of the design, no unmanned American Shuttle was orbited prior to its maiden flight (the Russian Buran shuttle did so, but the programme was cancelled, and no crew was launched). For Mars transit vehicles, system integrity and reliability, maintenance and repair, caution and warning systems and life support will have to be evaluated during the run up to dispatching the first operational crew to Mars.

It remains to be decided whether an Apollo-style step-by-step approach is to be adopted. One possibility is that elements of the surface vehicles, or a system for generating most of the consumables (including propellants for the return journey), will be dispatched to Mars before the human crew. Whatever might be the design and style of hardware, it will not resemble the 'Starship Enterprise', with a crew of 400, holodecks, spacious crew quarters, and a massive observation and recreation lounge. Nevertheless, the hardware will have to support considerable interactivity with the crew, together with facilities and amenities to support human lifestyles as near to normal as possible, both in transit and on the surface of Mars.

## HABITATION ON MARS

Habitation requirements on the surface of Mars will probably include the roles of lander and control room unless a separate vehicle is landed first. If an unmanned habitat is landed before the human crew arrives, then they will have to ensure that it is capable of supporting them before committing themselves to stay. They will also have to land nearby so that they can activate it and check it over, while considering the safety limits of their lander and EVA suits, before they have full habitation facilities.

Assuming a dual role for the habitation/control module, the area used for landing and take-off will effectively be the vehicle 'mission control' on the surface. It would be the master communications area and main systems area, and would probably include CCTV and direct viewing windows. It may even be the mission control area

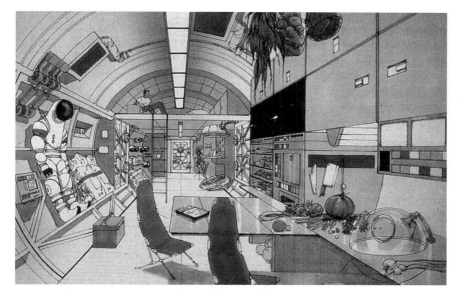

Although the final concept for a habitation module will include as many home comforts as possible, it will probably not resemble this artist's concept.

for telerobotic control and surface EVA operations. To gather all of these functions in one area is feasible with careful planning and application of technology. This is already the case with the Space Shuttle, on which the flight deck consists of the flight controls at the front (and duplicated at the rear), and the payload-handling gear located at the rear, together with monitoring equipment for the payload, viewing areas and communications.

On Mars, sleep stations will probably be used 'round robin' rather than individually. Most reports indicate a better work pattern when all the crew-members work together on a single shift, but if four to six people are on the surface they will need four to six sleeping berths, which increases the size of the vehicle and the volume of sleep stations. Unlike in flight, they will not be able to simply float in the middle of the room or strap themselves to the wall, due to the gravity. After being weightless for several months, they will also have to again become accustomed to lying down and using pillows.

A bathroom or, more practically, a shower (or sauna) would be needed – preferably a small and upright facility that could be stowed to save room. It would probably use recycled water for an all-over refreshing (and warm) wash. Depending upon the availability of water, the crew might have to use sponge baths for part of the week – but this would not be relished on a three-year mission.

On Skylab, the toilet was placed against the wall, but this will not be possible on the Mars lander. Its waste management system will be more like the Shuttle toilet, resembling a commercial aircraft toilet system, with probably more than one, and appropriate back-up facilities. Again, the thought of using 'Apollo-style' toilet bags during such a long mission would not be very appealing to even the most ardent space explorer, and it would introduce its own hygiene and storage problems.

The crew will also want a kitchen or meal preparation area – the galley. It can be quite small, as on the ISS, the Space Shuttle and Salyut, and it could double as the wardroom – a communal area where the crew could meet and talk through the day's events, experiences and impending operations. One of the more advanced designs was incorporated in the mock-up of the Freedom space station (c.1988), which was located at NASA JSC for several years. It was similar to a modern kitchen, with an area that protruded from the walls, like a large dining table. The surface could be adjusted in position for comfortable use (to avoid stomach cramps from maintaining position in microgravity), and there was an adjacent storage facility, TV screens, and provision for internet connections.

A wardroom would also be required – a rest area and off-duty lounge, away from operational considerations, in which to relax, watch a DVD, read a book or play a game. It could also be used for solitude, or to communicate with family on Earth (via Earth mission control). The Russians learned that there are some aspects of a mission that the crew do not want filmed or photographed, and even in this era of reality TV shows, the Mars crew will not want to be filmed using the toilet, resting, sleeping, or talking to their families. By the time the first crews depart for Mars, we should have learned enough about working in space that detailed work-motion studies will not be required, and while monitoring of operational activities will continue as part of the health care and mission documentation objectives, and the surface EVAs will be recorded for post-EVA debriefing, some parts of the day will not need to be studied in detail, if at all.

One major safety requirement will be for a refuge from solar flares – the effects of which can last for several days. There will be such a shelter in the mother-craft for use to and from Mars, and there will need to be another on the surface of Mars. It will be a place that the crew can retreat to during intense solar activity, and will also be used as a sleeping area or for storing food and items that could be damaged by radiation. However, the entire living area might be made a refuge, so that the crew will know that they will be protected during the time spent living on Mars. EVA and science operations could be conducted by robots during this phase, as the crew remains in the pressurised area until the danger has passed and radiation levels decrease.

Much will depend on the design of the vehicle and its mass. It is interesting to consider how the Lunar Module changed (although it was intended only for a short mission). It was originally a smooth and almost aerodynamic shape – but it was never intended to operate in an atmosphere, and so the designers modified it to save weight, producing an ungainly, spider-like appearance. At first the crew did not have a ladder to the reach the surface, and would have had to use a rope ladder and harness system; but simulations demonstrated the impracticality of this method – especially when re-entering the module at the end of the EVA – and so a ladder was added. The seats were removed after it was realised that the crew would not 'sit' in a $\frac{1}{6}$ -g environment or during orbital flight, and a 'stand-up' harness system was therefore incorporated.

During early planning for post-Apollo operations, Grumman designers realised that future missions might be more extended – perhaps two weeks or longer – and some designs of future LMs featured a sheltered laboratory called SheLab, or habitation modules and even ferry and logistic vehicles called LM taxis. Here, the ascent stage of the initially unmanned Lunar Module featured a pressurised module like a Spacelab laboratory or a Skylab wardroom in a smaller area. The crew could retreat into these facilities and take off their spacesuits, and live in it, while the LM ascent vehicle would be parked about 200 metres away. LM Taxis were also fitted with a platform on the descent stage, on which to place large rovers. This variation of the LM design was very impressive, and even featured orbital vehicles that were not designed to land on the Moon. It was the genesis of what became the Apollo Telescope Mount on Skylab, and though none of these manned LM variants reached production lines, the adaptability of the LM from its primary design demonstrates how hardware can be adapted for other missions. The variants of Soyuz that include Progress unmanned resupply craft is another example, and the same will be true for Mars surface (and possibly orbital) modules: a cost effective and practical use of proven equipment.

**The working environment**
The benefit of working in low Earth orbit is the lack of gravity and the extra volume. Even though the individual modules or decks are small, they are still very voluminous. On the Shuttle, underneath the flight deck, there are many facilities in a small area. It has proven useful, for a short mission, to fit a kitchen, a bathroom, a bedroom, a laboratory and storage area into a small volume, but this is not suitable for a very long flight. However, even for short missions there is still an issue with adequate volume in the crew compartment. Spacelab was a pure science laboratory, but when the Spacehab augmentation module was incorporated there was space for storage of equipment when it was not being used (although it proved difficult to find a particular item).

On Salyut, the Russians had problems in identifying 'up' and 'down', and so the 'floors' were made dark and the 'walls' made lighter – a psychological boost. In microgravity, lockers are very difficult to use if there are no restraints. Foam fittings were tried, but they occupied too much space, some of the catches proved to be a nuisance, and the crews had to jam the lockers and tape them shut. Another advantage, therefore was the inclusion of more strapping and other restraints.

Another problem which arises on large-volume spacecraft is a person's 'redundant' limbs. Someone floating in space is not like Superman, 'flying' head first. This technique is effective through tight spaces, tunnels, and from one end of a spacecraft to the other, but in reality the astronauts tend to orientate themselves to the room, even though they are floating above the floor. When moving through hatchways, however, they are dragging their legs, which can result in bruises if they forget to raise them. The rims of hatchways are now fitted with foam rubber, although fingers can be trapped when large items are being pushed through.

Reaccustomising to gravity has created difficulties for astronauts returning from long-duration flights in Earth orbit. Coordination is lost for a few days, and they will, for example, walk into a wall when intending to walk through a door. On arrival at Mars the crew will face the same difficulties, but in $\frac{1}{3}$ Earth-gravity after 6–9 months in space. Equally, after being on the surface for twelve months they will have to readjust to being in the vehicle that will take them home. Compatibility in vehicle design would be very sensible, as readjustment would be much easier. The controls for the Apollo Command Module and the Lunar Module were completely different due to their different roles, and so the astronauts had to adjust to a different layout. The Mars lander will probably become an initial habitat on early missions, so familiarity from similar control systems on the mother-craft will help the crew work productively and will reduce the possibility of potentially fatal mistakes.

The above highlights the considerations for working conditions, aptitude and training – broad brushstrokes that will help to define the parameters for living the Mars mission. However, consideration of the fine detail of human requirements reveals the enormity of such a project and the vast range of interconnected elements that will need to be in place to make the first human mission to Mars both feasible and survivable.

### Life support systems

During space missions, temperature variations can run to extremes, even in Earth orbit, with the spacecraft enduring sixteen sunrises and sunsets in a given 'day', and the consequent exposure to heat and cold, unchecked by atmospheric protection. In such situations, the hardware is pressed to maintain suitable living conditions.

If the air-conditioning on the Shuttle breaks down, the craft becomes extremely hot – like August in Houston. On Mir, however, if solar lock was lost it could become extremely cold. Keeping the temperature under control is important for crew health and comfort and their ability to work. When Apollo 13 suffered its systems failures as a result of an explosion, the crew soon had to contend with extreme cold and dirty air – but at least they were only a few days from home. On the way to Mars, such conditions would almost certainly be fatal.

With all the equipment running, and exposure to direct sunlight, the internal volume will become warm, and so the crew will utilise some of the heat. Some of it could be recycled, and the rest emitted into space. Equally, the spacecraft has to be kept cool, and so a balance must be maintained. On the flight to and from Mars, 'barbeque' mode – rolling the spacecraft to expose different areas to direct sunlight – could be used, but it would need to be accurate. On Soyuz 9 it was carried out a little too quickly. The cosmonauts felt rather sick, and the constant rolling affected star-sightings. On Mars, heat rejection requirements may result in the relocation of surface equipment and instruments to outside a landing zone exclusion area. Mars' atmosphere is subject to conduction, radiation and convection, but it is not known whether heat rejection will affect the equipment, and more importantly, whether it will contaminate the atmosphere and affect research in exobiology and other sciences.

In microgravity, unrestrained items float, and there is always the possibility that small objects missed during checks on the ground will unexpectedly appear. Very small items and dust have the potential to enter the astronaut's eyes or airways and

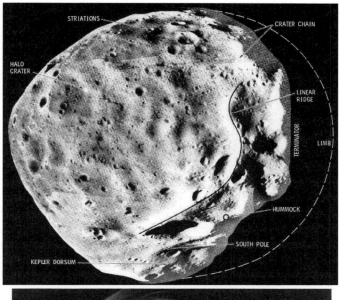

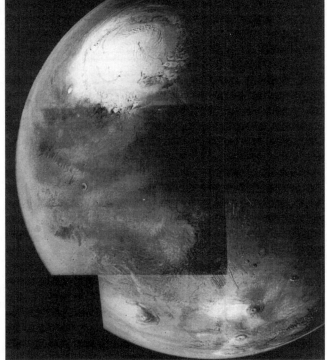

Phobos and the polar regions of Mars could be used as water sources during manned missions. (Courtesy NASA/JPL/British Interplanetary Society.)

to interfere with instruments, and so the astronauts frequently vacuum the air and clean the filters.

Leakages are also a concern, because the artificial environment inside a spacecraft can be easily contaminated. During a spacewalk on STS-98, one of the spacesuits leaked, and Robert Curbeam had to 'sunbathe' outside the spacecraft to evaporate some of the freon and then brush himself down. Inside the spacecraft the crew reduced the pressure of the air and wore masks for a few hours, but they could not detect any contaminants. On the surface of Mars, such measures may even be routine in order to keep the habitation areas clean. This will be important not only to avoid contamination on the surface, but also to prevent clouds of dust suddenly floating throughout the spacecraft after returning to microgravity.

On Mars, recycling systems will be needed for growing plants, using waste given out by humans and the lander systems. On Mir, urine was recycled to produce drinking water – but no-one had the courage to drink it. (The experiments continue on the ISS.) However, recycled urine could be used to provide water for other 'systems' – but it is never a perfectly closed loop, and it is not known how much air and water would be lost.

**The working day**
On Earth we are accustomed to working cycles of 8-, 10- or 12-hour shifts, day shifts, night work and weekend work. All of this is linked to time zones on Earth and the day/night cycle. In Earth orbit, crews use GMT on the ISS, Moscow time on the Russian craft and Houston time on the American craft. On the missions to Mars, however, Earth time will have little relevance, except for the psychology of the crew. To maintain a link with home, mission time may well be based on GMT, or 'Mars Mission Time' (Mission Elapsed Time). On unmanned Mars landing missions, the Sol (1 martian day) is used, and this might be used during human flights.

Whichever method of keeping time is used, some planning will need to go into the working day, even if it is not accurate to the last second as on Apollo. On Skylab it was found that exercise followed by a shower and a meal was not an ideal routine, and the schedule was therefore readjusted to make better use of the available time. On a mission to Mars, the crew may well not have much to do during the 6–9-month journey (as with the trans-lunar and trans-Earth coasts during Apollo), but they will still need to plan activities that will not upset circadian rhythms.

On space station missions (particularly Skylab) it was found that planning was best carried out during mealtimes, because it was quiet – although the crew did not always eat their meals together. On Skylab, at the end of each day, the Commander would read down to Mission Control how much food they had consumed, how much they had rejected, how much waste they had expelled, and all the equipment problems. They would also review the plan for the next day, and discuss what they had not been able to do. This inventory will probably be completed on the Mars mission – particularly so that the medical condition of the crew can be monitored, and dietary or workload adjustments can be introduced.

## Medical considerations

While the positions of the planets and the availability of propellants may still prevent an early return from Mars, the health of the crew will affect their work patterns, their ability to cope with the unexpected, and their mental wellbeing.

Injury or ailments will pose a problem on a long flight to Mars, as it will not be possible to immediately return a crew-member to Earth. The Skylab crews were taught how to pull teeth, carry out minor surgery, repair bones, and deal with many other medical requirements. During Apollo, an astronaut might have fallen over, broken a bone or be otherwise incapacitated and unable to return to the Lunar Module; and during the Salyut programme one of the cosmonauts had to be brought back to Earth because he was seriously ill. At Mars, the crew could potentially have to deal with all these situations, and their medical training and the facilities available will have to be extremely good. The inclusion of a sick-bay will be dependent on the design of the spacecraft, and even if the lander does not have a sick-bay there will certainly be a comprehensive array of medical equipment which would be familiar in a field hospital.

If the spacecraft has two or more decks there will be more potenial for the astronauts to fall, cut themselves, stub their toes, or break a limb, and they will need to be able to deal with such occurrences. It could have a major impact on operational procedures. If someone fractures a bone or pulls a muscle it prevents efficient working for several days, and another crew-member will need to take over the work; and if the patient needs constant attention, then two crew-members will be removed from the work routine, which will place additional pressure on the rest of the crew. Moreover, the situation would be far more serious if someone were to be injured during EVA – especially if the victim had broken bones or was suffering internal bleeding.

The most serious and tragic situation would, of course, be if an astronaut were to die or be killed. So far, no-one has died in space (all deaths have been within the atmosphere), but we will eventually have to address the morals and ethics of 'burying' someone in space because it is impractical or dangerous to bring them back. The Mars spacecraft and lander will have to have the facilities to deal with major illness or death on Mars, and body bags might be carried so that the deceased can be returned to Earth – chiefly for the sake of the family. (There is also the morbid question of whether a body in space deteriorates in the same way as on Earth.)

Then there are the familiar ailments – not measles, mumps or chicken pox, because the crew would have been isolated during preparations for the mission – but upset stomachs, colds, allergic reactions, and cuts and bruises. The crew will have shots, a record of medical history, and communication (hopefully) with Earth, and will be comparatively germ-free – but there is always the possibility that something will cause an infection. If this happens, then part of the spacecraft – perhaps the recreation or sleeping area – may also double as a medical isolation facility, separate from the operational area. Here, the medical kit would be stored, and there will certainly be medically trained staff, or perhaps specialist doctors, on the flight crew, but depending on the distance from Earth, more important operations may not be feasible, although operations may be less difficult in the low-gravity environment. The mission might even be equipped with an electronic library of training updates, refresher courses, drills and emergency procedures.

The medical officer will monitor the health and fitness of the crew (probably via a personal fitness programme), ensure that they are supplied with necessary dietary requirements such as vitamins, and submit them to regular tests (of their blood, saliva and urine, for example). They will also undergo psychological tests for determining their physical and mental state – including their thoughts on the progress of the mission. All of this information will be recorded to provide a database of human bodily functions and operations during a long, isolated mission.

Mission safety rules are akin to health and safety rules on Earth, where dangerous substances and high-risk practices are cause for concern. On Mars, this might include the handling of contaminants, and protective clothing, equipment, procedures and training would be necessary. NASA procedures are replete with guidelines and limits – such as Spacecraft Maximum Allowable Concentration (SMAC). Animal research facilities have been flown in space, but have sometimes proved troublesome. In 1985, Spacelab-3 carried rodents and primates in holding facilities, and their faeces leaked into the spacecraft. On a mission to Mars, health and safety measures will need to be particularly stringent. If a member of the crew cuts a finger, and martian dust enters the wound, what would be the consequences? As yet, no-one knows.

## Hygiene

As mentioned above, astronauts have to keep their spacecraft clean – chiefly with vacuum cleaners and antiseptic wipes. In microgravity, everything (if not secured) floats – including the crew, who can encounter debris, and have occasionally had it enter their eyes and lungs. Spacecraft require an airflow that allows particles to translate to filters – which in microgravity is easy to accomplish. On Mars, however, it will be as difficult as it was on the Moon. Even with gravity, the lighter particles may well still float in the air, and the astronauts will have to clean the spacecraft rigorously. This will have an effect on its design. If the astronauts have to spend a significant portion of their day 'moving the furniture around', they will have less time and less energy for other tasks. After an astronaut has had a shower (with the possibility of recycling the water) in microgravity, a vacuum cleaner can be used for drying and for clearing up excess water. But no matter how clean the environment, it requires constant attention.

On Mars, the crew might be able to have a bath, but such a facility would occupy more space than a shower; and as the bath would not be in continuous use, the space would be wasted for some portion of the day. Showering is more practical because it is quicker, it occupies less volume, and the water can be more easily recycled. On Skylab, the astronauts very much enjoyed a hot shower in microgravity, but it was very time-consuming. They could not stand under the shower and expect the water to fall, so they had to either hold the shower head against the body or force the water out under sufficient pressure. Cosmonauts on Salyut also found that they had to be careful not to inhale particles of water, so they wore a mask (which is rather unusual). The showers had to be kept clean to prevent the growth of mould. The Skylab astronauts also found that when they opened the shower curtain, the temperature inside the spacecraft was much lower than inside the shower 'tube', so they had to quickly retrieve a towel and dry off. On Mir, the cosmonauts eventually converted the shower into a sauna.

One of the first things that astronauts notice on visiting a space station is the stale and musty odour. On a Mars mission, the crew will be together for up to three years in an artificial environment, and no matter how much they clean the spacecraft, it will begin to smell. This is one of the seemingly small human habitability issues that will have to be addressed, together with tissues and toilet paper, wet wipes, and toothpaste, deodorant, soap, washing powder, and washing-up facilities. For a mission lasting such a long time, supply and storage will be a major consideration.

Humans exhale carbon dioxide – which, although not poisonous, is suffocating, as it prevents oxygen inhalation. It therefore has to be removed (scrubbed) from the air in the spacecraft. Lithium hydroxide has so far been successfully used as a scrubber on every human space mission, but its disadvantage is that many canisters have to be carried. In Earth orbit, space stations can be resupplied by Progress ferry or the Shuttle, but on a mission to Mars, everything would need to be carried and stored. In some systems, carbon dioxide is removed with zeolites – tubs or containers coated with a particular type of chemical which does not use any consumables. These can be connected to a vacuum and heated to dump all the carbon dioxide overboard, and can be used indefinitely. On Mars, carbon dioxide produced in the habitation and laboratory modules would be dumped into the atmosphere outside. However, plants absorb carbon dioxide and produce oxygen. Lithium hydroxide canisters could therefore be cycled via the closed environment life support system, providing the systems are compatible.

Personal hygiene and grooming will be an important issue. The crew will not necessarily need items such as make-up, but skin-care products, for example, will be available because of the dryness of the habitat's atmosphere. Equally, the crew will need hair-cutting equipment (unless they want three years' growth). This raises the issue of the removal of cut hair and whiskers, flakes of skin (the primary origin of domestic dust), nail clippings, mucus, vomit and other bodily waste. Moreover, each crew-member might not necessarily want or be able to use the same products or methods of grooming. For example, if there is a crew of six or seven (men), then the same number of shaving kits might be required. Some will prefer to use a wet razor, some will prefer an electrical razor, and some will grow a beard. In addition, there has to be allowance for allergies and other medical conditions and for the different genders.

**Clothing**

No crew-member will use the same spacesuit for three years, and there will probably be a selection of suits for different purposes. These might include soft suits for wearing inside the vehicle, for mobility, and for protection against the effects of leaks or contamination. There will also be EVA suits like those used on the ISS (Russian Orlan and American EMU), but including variants for use in orbit or on the surface. They will be various sizes, to fit different crew-members, and some will be adaptable, to allow more than one crew-member to wear the same suit, and to minimise the total mass and the amount of storage space. This would be necessary for contingency situations or emergencies. It will be of great disadvantage if the suits fit only certain people (as discovered on Mir and on the ISS), and so it will be important to have spare suits, adaptable suits, and the capability of repairing them.

Everyday work clothes will probably resemble the Penguin load-bearing suits, or the Chibis LBNP suits (which have always proved their worth). Shorts, slacks and T-shirts are also useful, but they produce particles and wear out. The crew will not use the same underwear for three years, and will require a plentiful supply. Clothes will be laundered by a steam process or similar facility. An alternative approach is that clothing would be disposable, as unless it was very lightweight and biodegradable there would not be sufficient storage space. It needs to be decided, therefore, how much clothing should be taken.

In space, the body's extremities – particularly the feet and legs – can become cold very quickly, and so the astronauts wear thick socks. During microgravity flights (especially on the Shuttle) the crews do not wear shoes, because they do not walk on the floor. Instead, they wear booties with thick soles – sometimes with Velcro pads to lock them in place at a workstation. With foot loops, the astronaut curls the toes to remain in place, although this is uncomfortable over long periods. This has led to the development of portable and adjustable restraint systems that grip the feet and legs more comfortably. On Mars, footwear will be required in the habitation module as well as for EVA.

For the lander or science laboratory, disposable clothing will be required for protection against possible contamination; and if particles of martian regolith tend to move from the airlock to the rest of the module, the crew might wear clothes which could be used for one day and then be discarded or recycled.

### Food preparation and meals

On short missions in the early days of spaceflight, food consisted of bite-sized bars or was squeezed out of a tube – and it was bland and unappetising. Fruit would occasionally be taken, but after the first day or so it began to deteriorate. Some of the early Russian cosmonauts were given calorie-controlled food, although sweets and desserts were also included. However, the food was not always suitable for particular individuals. Valentina Tereshkova, for example, did not want to eat one type of food because it made her feel ill – which caused some concern, as it was part of a controlled diet.

As experience grew, menus improved. They began to include freeze-dried, rehydratable food, and food warmers were provided for their preparation. On the Apollo missions, hot coffee was available, because hot water was a by-product of the fuel cells. Cans were sealed, and ring pulls were invented. These and other techniques evolved into today's familiar packaged food.

On Skylab, the astronauts selected their menu to repeat over a cycle of 5–7 days – although after a while they began to wish they had more variety. On these earlier flights, Russian cosmonauts and American astronauts were learning how to live in space, and were monitoring the digestive system. They had to keep a record of everything that they did not eat, and measure everything that they took into their body and everything that came out (from both ends). One setback for Skylab was that there was no capacity for an unmanned vessel or visiting crew to resupply the resident crew, and so the third crew had to take more food with them in the Command Module. Storage space on Skylab was gradually being filled, so energy bars were taken to supplement their main food – and at least one of the astronauts has said: 'I have never eaten a food bar since.'

The menu on the Shuttle includes about 120 items from around the world – biscuits, cakes, sweets, tortillas, curries, fruit ... and many other dishes, hot and cold. There are three meals a day: breakfast, lunch and dinner. One or two of the astronauts prepare the food for a day, and this duty rotates throughout the mission. Most of the crews say that they try to eat together, because it is the time during which they relax and spend time to talk about mission progress, problems, and private matters – or to simply look out of the window. Currency is not flown on missions, but astronauts have been known to barter food – usually for other food, but occasionally for 'rental' of a CD or video.

On the Shuttle there are three meals a day per person, plus the pantry items and emergency supplies; on space stations, fresh food can be supplied by visiting spacecraft; and on the Moon – which is only three days from Earth – monthly resupply would be practical. However, for a crew of six to eight on a three-year mission to Mars, it will be necessary to take large supplies of food (although it could also be grown during the mission). Moreover, even more food would be required in the event of an abort to Mars, as the crew would have to remain there until the arrival of a rescue craft.

Many astronauts like fresh food. This was not really an option on the early missions, because it occupied valuable space and was difficult to keep fresh. On the Shuttle, however, bananas and apples are taken for consumption during the first few days. On space station missions, ground crews store the treats at the back of the Progress resupply vehicle, otherwise the station crew find them and delay the unloading. This will not happen during a journey to Mars.

The habitation module on Mars will need to be more functional, because, after so long in microgravity, the crew must not forget that they are again in a gravity environment. They will have to be careful not to drop or drip anything into electronic or mechanical components, and will need to be careful how they handle the food – particularly with concerns about dust (and crumb) contamination. And they will again begin to use all the cutlery – which they will not have used in space. When they reach Mars they will not have fresh food unless they grow it, but consideration might be given to including a freezer, even if it is intended for nutritional or agricultural experiments. (On some missions a chill-box intended for scientific experiments has been used as a refrigerator – even for ice cream!) There will be supplies for three meals per person per day for the duration of the flight, plus contingency, and there will also be emergency supplies should the return flight be delayed. Before they left Earth they will have had food-tasting sessions. However, astronauts have said that food consumed in space does not taste the same as on Earth, because the taste buds react differently, and the flavours are more bland. Hot sauces, peppers, chillies, onions, and other spicy food awakens their taste buds, although they have to be careful not to upset their digestive system (because the toilets do not always work efficiently).

On the Shuttle and the ISS there is also a pantry for storage of snacks – usually high-protein food such as peanuts, chocolates, sweets, chew-bars and fruit-bars, which on short missions are no longer recorded in detail. Equally, they have taken food which, on the flight, they did not like. On Thomas Reiter's Mir mission, they had a Russian fish dish which was very salty. Everyone hated it, and did not eat it. This might also be the case on the Mars mission, as the crew faces the same food over and over again. They will have to balance what will be eaten and what might not be eaten. If there are scores of days-worth of such dishes, the crew might not want to eat it; but they will need to monitor their diet and calories – especially when they arrive on the surface of Mars and have to cope with the extra exertion due to gravity. The EVA workload will cause them to perspire a lot in their pressure suits, so they will want to replace the fluid loss with fitness drinks and vitamins. All of this will have to be taken into consideration for the food issue, and dieticians will play an important part in deciding the menu for the mission.

What will *not* be available is a push-button system producing tablets which hydrate into a gourmet three-course meal with a vintage wine. Food on the Mars mission will be as close to real food as possible, because it will need to be appetising, and the crew has to want to eat it. The challenge is that it has to be *enjoyable*. On the ground-based Skylab Medical Experiments Altitude Test held in Houston in 1972, Bill Thornton (who is a gourmet cook), found the food very boring. When the test was over (including the few days when the diet had to be continued), the first thing he did was to eat a huge T-bone steak. It is not exactly gourmet food, but he wanted something that looked, tasted and smelled like real food – and that was only after about seventy days! If a crew has had to endure a bland menu during an entire mission to Mars, they might find even the worst junk food appetising when they return to Earth!

**Waste disposal**
Once a meal has been eaten, there is the question of where to put the packaging and any uneaten food? This is a significant problem, which also applies to experiment wastage, broken equipment, used toiletries, packaging and, of course, human waste. On the early missions it was easy to dispose of food packaging. After the astronauts took out the food, they rolled up the containers to make them much smaller. They could also put the urine and faeces bags into empty storage containers. This was acceptable for such short missions, but became more of a problem as the missions became longer. On current space stations missions, rubbish is loaded onto the Progress supply vehicle, which is destroyed during re-entry into Earth's atmosphere; while the Shuttle can bring back rubbish and unwanted items of equipment for disposal on Earth. Excess water and urine is usually dumped overboard. This is spectacular, as it produces countless glittering points of light – but it can interfere with star-sightings.

On Skylab, the former oxygen tank in the S-IVB workshop was used as a trash airlock, and bags of waste, clothes, and all unwanted items were pushed into it. By the time of the third mission, however, it was difficult to close, as it was full of bags and there was a build-up of gases. If Skylab had continued to be used with more crews there would have been greater problems with rubbish – and this was one of the reasons for not mounting any further missions to the station.

During a three-year mission to Mars, a very large amount of rubbish will accumulate. The spacecraft will need to incorporate a trash airlock, but it will then effectively become a large dustbin – and what will be done with the rubbish after the return to Earth? Should a trash compactor be built into the spacecraft, or should fully recyclable materials be used? Trash compactors have been tried on long-duration Shuttle missions, but the rubbish still accumulates. On Apollo missions, urine bags were dumped on the lunar surface, as was unwanted equipment (although water and urine can be dumped in space). This raises perhaps the most sensitive point – whether the Mars landing site should be used as a dumping ground. On multiple EVAs, a small amount of air will be lost during evacuation of the airlock, adding to environmental concerns.

It is as controversial as the methods used at the early Antarctic bases, where rubbish was simply dumped. Now, however, a considerable amount of effort is being put into cleaning up these bases and in changing the attitude of those working in the Antarctic. They cannot dump things in the open, and have to recycle as much as possible. If bags of faeces or urine are dumped outside the facilities, the bacteria might affect the local environment – including the flora and fauna – and could interfere with biological research. On Mars, too, it would affect biological research (but not flora and fauna of course).

On Mars, urine would probably be recycled; but solid waste would pose a problem. In the early days, waste was collected in bags and taken back for analysis. In the Shuttle toilet system, the waste is drawn by an airflow into a system that distributes it around the storage facility. The astronauts call it the 'slinger' – because it slings the waste around the perimeter of a large vat, where it is exposed to a vacuum and freeze-dried. At the end of a 7–10-day mission it is disposed of using the same methods as used on commercial airline flights. For Mars, however, this is not feasible for a crew on a 2–3-year mission; but alternatives have been considered. On long-duration Shuttle missions, a compactor pushed the waste into an area where it could be be exposed to a vacuum and freeze-dried in a canister, which could be exchanged. But, as with lithium hydroxide, many canisters would be required for a mission to Mars. If a toilet breaks down there will need to be a back-up – which could still be the old fashioned toilet bags, although these are uncomfortable, and bring their own storage problems

On Russian space stations, and more recently on the ISS, waste is loaded into Progress freighters, and burns up during re-entry; but could it be ejected into space during transit to and from Mars? The bags would probably travel alongside the spacecraft; and besides, there is the question of whether this would be environmentally (or politically) friendly. It would certainly not be desirable to see it out the window for several months. On Mars there is no such option, as it would lead to surface contamination, but a storage area could be used. When the surface equipment and experiments are deployed, the space which they occupied in the lander would be freed up for use as storage; but again, contamination and hygiene would have to be considered. If a separate lander and habitation module were to be used on the mission, then some of the items could be stored in the lander while the crew uses the habitation module, and then transferred when the time comes to leave. The module could then be sealed and would have to be monitored for emissions, but it would not be environmentally acceptable over several landing sites.

### Sleep

Sleeping facilities are called 'sleep stations'. In space, of course, an astronaut can sleep while floating around – but he might bump into things, or wake up in another part of the spacecraft. Sleeping on the Moon was akin to a weekend camping trip. The crews could have a hammock hooked up between the telescope in the front of the Lunar Module and the back part of the Lunar Module, over the ascent engine cover, or could stretch it across the two flight stations. Sleeping on the floor was uncomfortable, because it was quite hard and very dusty.

Sleep stations on the way to and from Mars will be very similar to those on the ISS, because they have proven very efficient, although there may possibly be individual compartments because the crew will be using them for a long time. On Mars, there would ideally be a sleep station for each member of the crew, but on the surface they will not be able to float, nor use a vertical sleep station. The stations could be racked like bunks, but the astronauts will have to climb in and out of them by using a ladder – and when they wake up they will need to remember to use the ladder (at least for the first few days), as the effects of gravity might result in injury.

On the surface, an individual will probably feel the vibrations caused by others walking about or by the spacecraft systems. There will also be communications noise, rovers being deployed or delivering retrieved samples, and EVA astronauts exiting or returning through the airlock. Off-duty crew-members may well hear all of this, and might need to sleep away from the operations area, perhaps on the opposite side of the spacecraft.

The astronauts would no doubt prefer single, personal sleep stations, but the system might be similar to that used on submarines, with fewer bunks but a personal locker. On returning to duty, the astronaut closes his locker and changes places with the person coming off duty, who then opens his own locker. This minimises the number of bunks required, and the space saved could be used to carry extra propellant, food or equipment. This, of course, is a feature of full-time day and night operations, and it has been stated in several workshops on Mars activities that the crew would be more efficient following a simultaneous work/rest pattern – although this then increases the number of rest stations required.

On the surface the crew will need bedding, a mattress and blankets. They will also probably need a bedside light, access to the Internet, a personal stereo, a video (TV) screen, and an emergency system. Sleep will be one of the main areas for monitoring the wellbeing of the crew, and there may well be communications sessions so that mission control on Earth can determine their physical and mental state, how well they are sleeping, and how they are recovering from the exertions of the day. This will be particularly important if they have had to sleep in the pressurised rover or in a safe haven on the surface.

### Storage

In Mars studies there has been little mention of storage, but it is a major issue. One possibility now under discussion is inflatable modules – made of Kevlar, micrometeoroid-proof, and with little mass or volume until inflated with dry nitrogen, in space. Such a module could be used on the way to Mars as a portable storage facility for rubbish. The main issue is its disposal. When an object is ejected from a spacecraft it retains the same speed and vector until its orbit begins to decay. It can be sent towards or away from the Sun, but both options are difficult, and it would probably enter the same orbit as Earth or Mars. So, do we let it float around in space? Such objects might eventually burn up in a planetary atmosphere or in the Sun, but it is more of a moral point whether we are prepared to use space (or Mars) as a rubbish dump. The crew could use the inflatable module on Mars, but what would they do with it when they are due to come home – especially if it is full of faeces bags, urine bags and waste food containers? It is possible to dispose of items from the Antarctic bases, but it requires a great deal of money to do so. Bringing all the waste back from Mars may simply not be an economical or logistical option.

One option is to keep careful track of the inventory – what has been used, what space has been freed up, and what items need to have room put aside because they will be returned to Earth. Bar-coding will certainly be used, but such systems are useful only if the crew keeps the system updated. On several missions it was found that if items are not properly stored after use, the next crew-member will not find them, thus causing delays and disruption to the working day. However, the correct placement of toilet rolls will not always be uppermost in the minds of the crew!

To maintain a pristine environment the crew will need an air supply, water, sample storage facilities, and possibly storage 'vaults'. Documentary recording and the writing of reports will be part of the routine, and mission reports will be relayed to Earth throughout the mission. Now, of course, we have moved towards digital storage rather than photographic film, but certain film and still cameras are (at the moment) still used. It may well be that by the time of the first mission to Mars, photographic film will no longer be used; but digital media can be affected by cosmic rays, solar flares and other radiation, so there will need to be sheltered areas in which to keep the cameras, and possibly fast uploads of data to Earth to avoid loss of data and to free up the cameras. Pristine facilities will be required for storage of samples of Mars' soil, rocks and atmosphere for return to Earth. The care and operation of these 'vaults' will have to be carefully controlled so as not to harm original material.

### Communications

Besides official communications with Earth, the crew will want to claim some personal air-time for family affairs. It will be interesting to know how this will affect the crew, because of the distance and the time. The decision whether to send a bachelor or spinster astronaut, or a husband and wife, on such a long and distant mission will be difficult. Families will eventually fly in space, but probably not on the first flights to Mars. This will more probably happen when a research base is established and research teams complete a detachment to the base for a five-year tour, similar to military deployments on Earth.

The crew will certainly receive world, local and personal news on a regular basis, but will it be filtered before they receive it? In 2001, news of the attack on New York could not be withheld from the ISS crew because they flew over the city and saw the great plumes of smoke. On the Mars mission, however, mission control on Earth might not want to inform the crew of certain events that might affect their performance, personal health or harmony, although there will be mission updates both ways, such as information on solar storms, and changes at mission control. The crew commander will report on the events of the past few hours, and update the status of the mission accordingly. There will also be times when the crew will want to escape from work and not communicate with anyone.

Personal and crew events – such as birthdays and other celebrations – will be observed, and information on national and world events will be relayed to them – but it is not known how they will relate to all of this at such a great distance. Astronauts who have been away at Christmas, for example, have been able to open presents, have turkey lunches, and even imbibe brandy or champagne. With the Mars crew being away for three years they will celebrate birthdays – and perhaps, in certain lockers, they will find hidden presents. Equally, if there is an international crew (which is probable) they will have to have international celebrations for different cultures and faiths. A major sporting event such as the Olympic Games, the World Cup or the Superbowl, will probably be followed quite ardently, and there will be more personal matters to celebrate, such as anniversaries, childbirth, or graduation of children from university. The crew must also be able to vote in parliamentary or government elections. Some astronauts have very political interests because they intend to follow such a career when they return home, so they will certainly be keeping an eye on political programmes.

Contact between people in remote locations on Earth has now become much easier, due to e-mail and the Internet, and attached video or image files help people keep in touch with distant family members. Such communication with the Mars crew is probable, but it may well be filtered through mission control on Earth. The crew will not want to be swamped by e-mails, and care has to be taken to protect against computer viruses and hackers. One possibility is that a Mars-orbiting satellite could be used as an e-mail post-box.

Compatibility and teamwork is very important. The crew will be a small, isolated group of of humans, requiring high morale and constant motivation, and they will need to be able to cope with possible bad news from home – adverse world events, or perhaps a family bereavement. Individuals will have off-days, when they think that they have made a mistake or have failed, and the rest of the crew will have to help and remotivate them. They will not be home in a few days, and they will have to overcome such difficulties. They will also have to be available for public appearances, relaying TV and radio programmes to Earth.

The crew will probably evolve into a quite tight-knit community; and if there are two landing crews, or an orbiter crew and a lander crew, they could maintain morale through competitive spirit.

**Leisure time**

Space crews have often found that the best opportunity for free time is when everyone on the ground thinks they are asleep. They do not need as much sleep in space, and the best time to observe the Earth is after they say 'goodnight'. On the ISS, crew-members pass the time by looking out of the window at Earth while they are exercising; but during a voyage to Mars, the only noticeable change will be the apparent movement of the stars when the spacecraft enters a 'barbeque' roll. Any extra window in a spacecraft is an additional structural weak point, but for the crew it is an important feature for 'flying' the spacecraft safely and efficiently, and for scientific and recreational observations. The inclusion of a window in the Mercury spacecraft was an early contentious issue for the first American astronauts and spacecraft designers. The martian landscape is barren and desolate, but it can look spectacular, although the crew would need to become accustomed to a very unfamiliar colour scheme – pink skies, and an orange or red surface – and the effects of the weather.

As on current longer missions, Mars crew-members will not work every day. Many astronauts on the shorter missions are under so much pressure that the days pass quickly, but on longer missions the difficulty is in maintaining consistency. There is the possibility that an astronaut, 'stuck in a can' with few windows and the

Although most of the surface operations will feature scientific or engineering goals, there will be time for recreational activities and ceremonial EVAs, such as visiting the sites of earlier unmanned landings. (Courtesy Pat Rawlings.)

same people, might grow to dislike the other crew-members – or there might be a disagreement so that they find it necessary to work at opposite ends of the spacecraft. This has certainly happened on the Russian programme. This, again, raises the subject of crew compatibility. They might live together famously, but there will still be times when they want to talk to someone else – perhaps a celebrity on Earth. During early Russian space station long-duration missions, ground controllers invited 'guests' to talk with the crew in space. These included artists of stage and screen, singers and musicians, world leaders, and other famous individuals.

The common factor in all human spaceflights thus far – including the Apollo lunar missions – is that the astronauts and cosmonauts (and the taikonaut) could see the Earth. During the journey to Mars, however, this will be extremely difficult, so even on a day off the crew might be tempted to carry on working – not only because of there not being much else to do, but also to help take their minds off what they are missing on Earth. They will also need to take breaks from the routine to keep themselves fresh, and will need other activities to occupy their minds. There will be sports and games, books, music, and so on. (Chess has already been played in space, as has darts – with Velcro tips.) At Mars there will be the possibility of, for example, multiplayer interactive computer games between the orbiter crew and the lander crew.

Recreational activities might even extend to outside on the surface. On Apollo 17, Jack Schmitt talked about skiing on the surface of the Moon, and Apollo 16's rover was used for a 'lunar Grand Prix'. Golf has been played on the Moon, and items have been thrown on the Moon – so it is not inconceivable that such activities will also take place on Mars. The astronauts on Skylab conducted gymnastic exercises around the inner walls in the microgravity environment. 'Sport' has even been undertaken in isolated places on Earth. One man in the Antarctic uses a kite, rather than dogs and sledges, to transport himself and his equipment. Will kiting become popular on Mars?

One of the ISS crew – Don Pettit – is an amateur astronomer, and during his extended flight he used his leisure time to carry out observations. This could be a leisure activity on the flight to Mars – but not when the spacecraft is in a 'barbeque' roll. On Mars, the areas of interest (assuming that leisure time will be available) would be the same as those followed on the ISS: amateur science – especially for educational purposes. People on Earth could be involved with the progress of the mission, amateur experiments, and presentations about living and working on Mars.

Another option for the crew-members would be to pursue a PhD or MSc degree or other qualification, or they could take the time to write a report, a book or a diary, or conduct teacher-in-space experiments or small personal-interest experiments. On Mars, their geological skills would be improved, and student experiments would sustain interest during the long flight. Crews usually enjoy these types of activity, providing they do not interfere with health and safety and the operational aspects of the mission. On Skylab, for example, they ran 'Saturday Morning Science' shows. Jack Lousma was famous for his opening line, 'Hello Space Fans', which attracted younger viewers to the wonders of spaceflight. The same might be true for the Mars crew – humanising the adventure and providing a real-time outreach link with Earth. This might also help prevent the possibility of the public becoming bored with it (as they did with Apollo until the Apollo 13 accident). It will, after all, be a long flight across emptiness, and there will be nothing much to see or report about except at each end of the journey.

An additional leisure interest on Mars would be art – either traditional or digital. Sketches and colour notes could be translated to a finished painting after the return to Earth; but if there were sufficient time, the paintings could be created and completed on Mars. Before the mid-nineteenth century (when cameras first became available), paintings and drawings formed a very important part of the records of an expedition. For example, in 1772 William Hodges accompanied Captain Cook on his second voyage, and was the first to paint pictures of what we now know as the Antarctic continent. Pictures such as these were circulated when explorers returned. Drawings were fleshed out into paintings and incorporated in lecture tours, and were also published in journals, newspapers and magazines, to build interest and to raise public support for another expedition. Cosmonaut Alexei Leonov and astronaut Alan Bean have painted scenes incorporating the sights and 'feelings' of being in Earth orbit or on the Moon,[1] and artist Andrei Sokolov has sent paintings to Soviet space stations and asked for the cosmonauts' criticisms and comments. In this era of digital photography, copies of drawings and paintings produced during the journey, or on Mars, could easily be sent to Earth. There might even be an artist's electronic palette and easel – a concept used by leading space artist David A. Hardy in his novel, *Aurora* – a science fiction story set on Mars.

There will also be poetry and other artistic pursuits such as music. Musical instruments have already been taken into space, and Talgat Musabayev and Pavel Popovich are wonderful singers. One of the crew-members might even learn to play an instrument on the way. On one of the long-duration ISS missions, Frank Culbertson took his trumpet into orbit with him. The Russians told him that the best place to play it was in one of the nodes, and they closed the hatch to prevent the noise from spreading through the rest of the spacecraft. However, Culbertson did not realise that the action of blowing the trumpet would change the air pressure – and he could not open the hatch. He was a little concerned that the Russians did not tell him about this – again, an issue of crew compatibility.

A particular difficulty will be homesickness. Even during missions in Earth orbit, the crews miss the sounds of Earth, such as birdsong, rippling water, waves breaking on a beach, leaves rustling, a baby crying, or their families' voices, so they take audio cassettes with them. This is acceptable when there is a clear view of Earth; but at Mars, a reminder of home could increase the feeling of loneliness and isolation. A linked issue here is the security of the crew-members' families. The astronauts will have to be assured that their families are being supported and cared for during their absence, and that they are being protected against unscrupulous forces and the media. They will also need to know, in the event of a major problem, that their families are being kept informed.

**Mechanics and repairs**
Emergency drills will be repeatedly practiced in order to maintain training and response times, and systems will be repeatedly tested. If a certain part of the spacecraft needs to be closed down, the crew will need to ensure that the appropriate

systems – such as hatches and seals – are in working order. Communications would also require contingency and back-up options. If the main communications systems break down, what would be used to keep the crews in touch with each other and with Earth? Perhaps a system similar to Morse code would be available as a back-up. In the event of a complete failure of communications, there would also need to be an electronic or paper copy of basic procedures for the crew to follow.

As with Apollo, some equipment – such as the lunar lift-off engine – cannot have redundancy. (The Apollo mission planners introduced great risk by relying on only one engine.) There may also be a different system, rather than a ladder, for moving to and from the surface. Some designs for martian landing craft have included a ladder of about ninety steps – which is obviously impractical, as it would require great effort to climb it after a long EVA or in the event of an emergency, and to fall from it would be extremely dangerous, even in $\frac{1}{3}$ g. It would be much more sensible to have the exit and the hatch as close as possible to the surface, and there could be a lift system or just a few steps.

The suits and EVA equipment will probably be supplied with an extensive repair kit – probably like those being developed for the ISS, where the station crew will be able to seal holes in the pressure vessel. There may also be items such as the tile repair kits being tried on the Shuttle, and a set of DIY tools. There will be a certain amount of damage and breakdowns – natural or accidental (but hopefully not intentional) – that will have to be remedied or repaired, possibly outside, and this will add to timelines and difficulties. These tasks would probably be carried out by EVA specialists – but the work might fall to a crew-member on his day off, which could affect the morale of that person, the scheduling of the workload and the sleep and rest patterns. On space stations at the moment about a third of the time is taken up with constant repair and maintenance, so such activities have the potential to be a major issue on Mars.

It is very telling that among the first items that Bill Shepherd added to the Zvezda module on the ISS was a workbench. The Skylab astronauts also found that they needed a work surface, a 'tool room', or a useful place to store frequently used equipment. Many astronauts and cosmonauts have tinkered with motor vehicles, aircraft or engines, or have occupied themselves with woodcraft, home improvements and similar hobbies, and so are accustomed to routine repairs.

On Mars, spare parts and replacements (including items to be cannibalised) will also be required, dependent on available storage space and expected breakdowns. It will also be useful to be able to make items from sticky tape, book covers, food containers and items of clothing, and materials such as thick plastic, thick card, masking tape, bulldog clips, corded ropes, bungee cords, and so on. No training is required – only common sense, ingenuity and skill.


## LEARNING TO LIVE IN SPACE

Humans first learned to survive in space during the pioneering programmes of Vostok, Mercury, Voskhod and Gemini; we learned to work in space onboard Soyuz and Apollo; and we learned to live in space onboard the Skylab and Salyut space stations. The skills of living *and* working in space developed with Mir and the Space Shuttle, and continue on the ISS; but we have not yet attempted a journey to Mars. All of these skills, and many new ones, will be required for completion of a safe, successful and productive mission. The first flights to Mars will confirm whether we really can venture far from Earth, perform useful work, gain great benefits for all on Earth, and, most importantly, survive and enjoy the adventure.

## REFERENCES

1  Alan Bean and Andrew Chaikin, Catalog of Paintings and Drawings.

# The return

The astronauts have been lucky enough to explore the surface of Mars for either a month or a year or so. They have collected copious amounts of regolith and rock samples, searched for evidence of life, and possibly examined some of the robotic hardware that has landed on Mars since the 1970s. With the benefit of initial analysis in a scientific laboratory module on Mars, the number of samples has been reduced to the finest examples, and the mission has proved that it is possible for humans to live and work on the Red Planet. Now the launch window for a craft to return home by the most efficient route has come around. Propellants for the flight home may well have been manufactured from local resources on Mars, and drinking water, or water to be used as protection against radiation, may have been 'mined' on Mars. Plants may have been grown on Mars and harvested for sustenance during the return journey, and the samples have been stored or launched into space for stowage in the mother-craft. It is time to return home.

## THE RETURN TO EARTH

During the 1997 Mars Surface Mission Workshop, consideration was given to the requirements prior to the departure of the crew from the surface, and some of these techniques will certainly be simulated and trained for during the time on the surface. They will be developed as the final mission profile is determined and, in particular, if any follow-on mission will include a landing near the first site. In this case, full or partial reactivation of hardware and facilities from the first landing may be part of the subsequent mission's profile.

### Pre-departure operations

For the first mission, therefore, clean-up and return preparations would include parking and storing surface mobility systems (and protecting them in 'hangars' if necessary), closing down non-essential systems and equipment, closing out laboratory functions and power supplies not required to maintain systems and equipment (remote science and weather stations, timed landing beacons, and so on), and shutting down or transferring command of remote vehicles from the ground to orbit or even to Earth-based control and monitoring.

For several weeks prior to departure, the habitation module will have to be cleaned and placed in dormant mode if it is to be reused. This may include leaving a 'time capsule' of material to be examined by the next crew, who will be able to evaluate any deterioration of the facility before committing themselves to their own stay on the surface. All systems will have to be turned off or placed in a dormant mode, with the life support systems not requiring full operation between crew visits. This type of decommissioning has already been performed on Russian Salyut and early Mir operations. The downlink of information received, recorded or awaiting dispatch to Earth may be accomplished at this time, supported by an inventory of hard-disc data taken to the ascent vehicle for further analysis during the return to Earth. In the event of a failure of the launch vehicle, the surface facilities will also need to be capable of rapid reactivation to enable the crew to survive pending repairs or rescue.

Verification of the ascent vehicle's systems will have to be completed, and throughout the surface stay an emergency procedures drill will have been regularly enacted to confirm operational readiness and to maintain the crew at peak performance levels. All returned cargo will have to be loaded and stowed in the return vehicle, and at the same time the lift-off mass and centre of gravity will have to be frequently recalculated to ensure that it falls within safety limits and the operational performance of the lift-off propulsion system. Crew transfer to the ascent vehicle will have to be completed, possibly in a staged operation (as with exchanges of ISS crews to the returning Shuttle vehicle), to configure the return vehicle early and to extend surface operations for as long as possible.

One of the most debated topics – and one that has raised most concern – is the problem of dealing with waste and rubbish (as previously detailed). Although recycling will be an important consideration in hardware design and crew training, it is inevitable that rubbish will accumulate. On Mars there will probably be no facilities to launch waste off the surface, and it would be impractical to return it to Earth for disposal – particularly after an eighteen-month stay. It has been suggested that an external shelter could be constructed to store containers of trash outside the habitation module. However, it will have to be properly constructed, and leak-proof, to prevent contamination of Mars, the habitation facilities and the samples.

**Coming home**
The choice of when to depart Mars for the return to Earth depends on the propulsion method for the parent craft. Using the engines available at the beginning of the twenty-first century, the craft can either stay at Mars for 1–2 months, or can opt for a longer stay of about eighteen months – depending on the alignment of Earth and Mars, and whether there will be a fly-by of Venus to bend the flight path and alter the velocity. A favourable line-up of Earth and Mars occurs once every twenty-six months, but the line-up launch window lasts for only a certain time, after which it is not worth the effort of chasing the other planet as it continues to orbit the Sun.

An artist's concept of a spacecraft beginning the long flight back to Earth. Performing the trans-Earth trajectory burn. (Courtesy Pat Rawlings.)

Once the return is underway, the crew will have to keep themselves fit, healthy and occupied during the long flight home. Apart from exercise and leisure activities (they may have exhausted their supply of books and films unless they have uploaded others from Earth, but at least they will have the fun of weightlessness again), they will probably have to apply themselves to work and experiments, to keep alert and active.

The martian samples could be partially analysed during the return journey – which would be justified if some of the crew-members were Principal Investigators for particular experiments or scientific disciplines. They would select which samples would be brought to Earth. It is their choice because it is their work. The scientist-astronauts would then be able to prepare papers based on their observations on Mars and early measurements made in the science laboratory and onboard the spacecraft during the homeward flight. Studies could not be as intensive as the investigations on Earth, but as the Principal Investigators they could carry out much of the initial work. Modifications and corrections could be sent to Earth via e-mail, and papers might be published in scientific journals while they are still travelling back from Mars, and they might even produce a Nobel prize-winning paper.[1] For example, the finishing touches to an article for *Scientific American* were submitted by scientist-astronaut Edward Lu while he was onboard the ISS;[2] and Gennady Padalka and Mike Fincke – also on the ISS – submitted a paper on remotely guided ultrasound examinations, which was published in *Radiology*. Other tasks during the return journey will include lengthy debriefings by staff on Earth, exercise (especially if there is no artificial gravity), maintenance, and solar monitoring. Debriefings could include early observations and guidance to pass on to a subsequent crew for a follow-up mission, or to adapt Earth-based simulators for the next crew's training programmes. Maintenance tasks will be required, as the mission will be up to two years old by the time of return. Monitoring of solar flares and coronal mass ejections would be a constant background task, and biomedical monitoring of the crew would be essential in establishing their fitness to return to Earth, and especially to monitor the development of any potential illnesses or health problems.

The crew might first land at a Moon-base – a 'half-way house' between weightlessness and 1-g on Earth – to provide them with the opportunity to readjust to some gravity after the long journey. After being in a closed environment for two years away from Earth, they will also be susceptible to Earth infections, and a lunar facility might be used as a quarantine station for the crew and the samples – a last check for possible signs of Mars-induced health conditions or potential contamination before they are introduced to Earth's environment.

### Public affairs at home

As the Mars explorers wend their way home, communication with Earth will become much easier, due to the shortening signal delay time. The crew could look at the Earth through a telescope on the spacecraft to see the disk growing larger, and to see the oceans and the clouds – a welcoming sight. There would be messages from Earth – from family, friends, journalists, politicians . . . a clamour of people wanting to talk to those who had been the first to walk on another planet. This would also be an indication of the gruelling public affairs programme that the crew would undertake once it was deemed safe for them to do so. The mission will have huge historical significance, and will signify humanity's initiation as a multi-planet species. For some of the crew, however, it could be their last spaceflight. Apart from the potential problems of radiation exposure and the long period away from Earth gravity, these crew-members would become valuable national and international ambassadors, in the same vein as Gagarin, Tereshkova and Armstrong. It may be that their respective countries decide that they are too important to risk them in space again.

## MARTIAN SAMPLES

As previously discussed, automated sample-return missions will precede human missions.[3] They will be aimed at surface material safety studies, and sample selection will be much more limited – but the challenge of safely returning samples and analysing them on Earth would be same as for human missions. A robotic sample-return mission would proceed as follows.

Martian samples are collected at a central site and transferred to a Mars Ascent Vehicle (MAV) for launch into orbit, where the MAV undergoes automated rendezvous and docking operations with another craft before the engine on that craft fires to leave Mars orbit en route for Earth. In one scenario, the exterior of the return craft would be catastrophically destroyed before arrival on Earth, to minimise the risk of contamination by the craft that transferred the samples in Mars orbit. The samples would be housed in a capsule, which would be released for re-entry and ultimately a parachute landing and collection at a remote, open location such as Woomera in Australia or the salt flats of Utah in the USA. The capsule would be grabbed in the air by aircraft or helicopter (hopefully without re-enacting the NASA Genesis capsule debacle in which samples of solar wind, collected for 850 days at the Earth–Sun Lagrange-1 point, crashed into Utah when the parasail recovery system on the capsule failed to open).

It has been proposed that martian samples should not be brought to the Earth's surface. They could be studied on a space station, for example, or on the Moon, using the same way station and quarantine facilities as the crew. A NASA study of an orbital facility, called Antaeus, has investigated samples like martian regolith/ rocks in isolation.[4] While such a facility would negate the advantages of bringing the samples back to Earth for detailed study, it would have a major advantage in public relations in that the samples would be safety out of sight, and the perception of risk would be considerably reduced.

**Analysis**

The best location for detailed analysis of martian samples is Earth, where the latest techniques and hardware would be available. Hardware that requires dangerous consumables or has a large floor footprint, computer requirements or power requirements would be impractical to launch to a space station or a lunar base, and would need to be back on Earth. Samples would be divided in clean laboratories and shipped around the world for simultaneous study by many research teams, using different equipment and techniques. This is how the lunar samples from the Apollo Moon landings were processed, but it will take about a decade to prepare the Mars sample receiving laboratory on Earth.

Once martian samples are released from the quarantine facility (they would first be sterilised to negate the possibility of any pathogens), analysis would be carried out on Earth, in a BioSafety Laboratory level 4 – a combination of a clean room (to prevent contamination from the Earth environment) and a quarantine facility (to protect the Earth's environment from any martian pathogens). This is the first facility of this type. (The laboratory in the Dustin Hoffman film *Outbreak* is like the Center for Disease Control and Prevention in Atlanta, Georgia. It is very good at containing an infectious disease but is inherently a 'dirty' environment in terms of contaminating the samples contained within.)

Back contamination is a major issue. The samples might carry an infection, microbes or viruses, which could affect the crew while on Mars or during the long flight from Mars to Earth. Any problems should manifest themselves on Mars or during the early part of the flight. Quarantine would not necessarily be needed in the same way as it was applied during the Apollo lunar missions. At that time, people were afraid that any infections from the lunar samples that might be carried by the astronauts coming back to Earth would only become apparent some days or weeks after they left the Moon's surface, because of the short distance and the mission duration. With Mars there would probably be signs of problems either on Mars itself or on the flight from Mars back to Earth, but the possibility of contamination or infection occurring only after exposure to Earth-like conditions cannot be discounted.

## Contamination

Contamination works both ways – either tainting martian material with that from Earth, to produce false indications of life, or uncontrollably releasing martian material into the wider Earth environment.[5] We are afraid of the possibility that martian samples could contaminate Earth – but the pristine samples might become contaminated by materials on the Earth. Keeping Mars pristine and keeping Earth safe are two major considerations for any return mission. The former is not just to keep the wilderness pristine, but to preserve the integrity of the search for life on Mars. If we find life, or even traces of life, was it brought to Mars with humans or robots? This is complicated by the fact that we do not really know how life started on Earth. It is possible that microbial life first evolved on Mars, because in the distant past, conditions may have been more clement there than on Earth, whether due to meteoritic bombardment levels or the brightness of the early Sun combined with the atmospheric composition and axial tilt of Mars. Life could have migrated to Earth on a meteorite, because meteoritic material is constantly travelling between the two worlds. Due to Earth's higher gravity, it is much more difficult for material to be 'knocked off' to travel to Mars, but it might be possible. Studies have been carried out (in Earth orbit) into the survival capabilities of microbes in open space – and they might be shielded inside a meteorite.

Keeping Earth safe is essential for public relations purposes at the very least. Due to films like that based on Michael Crichton's novel *The Andromeda Strain*,[6] people have been ingrained with a fear of life from space. There might be little possibility of life on Mars – especially life dangerous to humans – but maximum measures must be invoked to protect Earth. The last-minute addition of planetary protection against lunar rocks was rushed though by well-meaning scientists; but protection against martian rocks is complicated by the additional need to prevent contamination of the samples by material native to Earth.

Assuming all the appropriate checks and quarantines are passed, the returned samples would be cleared for scientific investigation around the world. Testing would include chemical analysis, microscopy, metabolic testing and an extensive range of other tests and experiments to obtain the most from the samples. Such tests would, in part, verify the initial analysis carried out on Mars or during the return journey, and there is enormous potential for scientific papers, reports and theories. It remains to be seen whether any of this will change our fundamental ideas about the formation of Mars, the Solar System or life on Earth, but what it may do is influence any subsequent missions to Mars and, in the long term, any decisions about how we choose to utilise the Red Planet in the future, if at all.

## THE RETURN TO MARS

So, the first human mission has landed on the planet Mars, stayed for months or years, and returned to Earth. Further missions may follow to continue the exploration, in a manner that only humans on the ground can accomplish. They may return to the original landing site in order to continue infrastructure development, but another possibility is that a series of uniquely different sites will be explored. Mars has a wide variety of terrains, and they all offer the tantalising possibility of evidence of past life, or even a surface niche for extant life. Despite being smaller than Earth, Mars has the same land surface area as Earth, so there will still be plenty of ground to cover after the first mission.

It all depends on the goal of the first mission – whether it is competition with other nations, the creation of an off-world colony, or something completely different. The most unlikely possibility is a purely scientific mission – but will the first mission be a 'one-shot' extravagant gesture, the first of a short series, or the beginning of something far bigger? This will depend on variables such as what is discovered by the first mission; the public, government and scientific reaction to the mission; and changes to international and national organisations or governments behind the mission

The first landing will have 'exploration' at its heart – implicitly or explicitly – searching for past or present life, and mapping resources that can be used for the expedition itself and then by later visitors. Science and applications notwithstanding, this is also an exercise for the human spirit – the intangible 'because it's there'.

Studies by scientists ideally require a base out of which to operate. It is a long way back to Earth – in both time and distance – for sample return, in order to carry out analysis, and research projects will need a turnaround more rapid than every two or more years. Such analysis will begin in the laboratory module during the first mission, and it could develop into continuous occupation of a laboratory module that remains on Mars, with a later crew arriving to relieve a crew already working there. This is the beginning of a martian base, and it might begin to grow at the very first human landing site. Plant-growth trials before, during or after the first landing will lead to full greenhouse construction, and an early version will help to sustain the stay time of the very first humans on Mars after their landing.

The supply line between Earth and Mars could be kept open using either robotic craft propelled by solar sails or by continuously cycling craft. The latter includes the VISIT design, continually travelling between the Earth and Mars, receiving fresh supplies and crew at Earth, unloading both as it approaches Mars, and exchanging them with the crew and samples already there to bring them home. The base on Mars would be analogous to Antarctic research bases, with its own laws and administration. Territory on Mars cannot be claimed by national governments, so bases would probably fall under international administration.

Permanent settlements on Mars would be controversial, and could lead to claims of colonisation – a contentious and divisive subject – but the advantage of a colony on Mars is that there are no native inhabitants to be subjected to oppression, enslavement or forcible relocation. Of course, this assumes that there are no local life forms, no matter how 'primitive'. An example of politics and reasoning behind a Mars base is graphically illustrated by the British 'Base A' – the first permanent Antarctic base, established on Weincke Island in the Antarctic Pensinsula, as part of Operation Tabarin in 1943–44, during the Second World War. It was ostensibly used for science, but the primary reasons why it was established were to deter German commerce raiders (such as U-boats) and to stake a British claim to the territory.[7] On Mars, there will not be a military war, but if sufficient mineral deposits are found to justify investment in mining and colonies, there may well be a commercial one.

## BEYOND MARS

The first human settlement on another world might be situated on the Moon, but otherwise it will probably be on Mars. After this, attention could turn to the asteroids and the moons of Jupiter and Saturn – but the outer Solar System is far more distant than Mars, and utilisation of local resources would be even more essential. Asteroid exploitation could be tested on Phobos and Deimos, and the

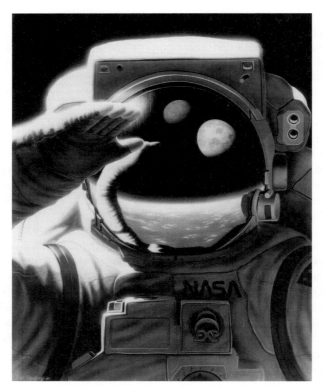

Within grasp. Our near-term future will see exploration and utilisation of Earth, the Moon and Mars. (Courtesy Pat Rawlings.)

development of protection against radiation would be paramount for colonies or mining operations on the moons of Jupiter, within the maelstrom of the Jovian radiation belts. This, too, may first be developed for use on Mars or on the long cruise to or from that world.

On reaching low Earth orbit, a traveller is 'halfway to anywhere' – the big leap out of Earth's gravity well. A journey to Mars takes a traveller to the edge of the inner Solar System – a significant step out of the Sun's gravity well. Within the orbit of Mars are the planets Mercury, Venus and Earth, the Moon, and, occasionally, a few asteroids – all basking in the solar wind and radiation. Beyond Mars lies the main asteroid belt, the gas giants Jupiter, Saturn, Uranus and Neptune, the small rocky world Pluto, the Edgeworth–Kuiper Belt, and, at a distance of about 1 light-year, the Oort Cloud of comets (believed to be the original building blocks of the Solar System). Could Mars become a stop-over and refuelling point for both inbound and outbound craft? If life is not found there, and if it is deemed too expensive or too dangerous to establish colonies, perhaps there could be (automated) mining for propellant production – in which case, could Mars become a deep-space 'garage'?

Terraforming Mars is a controversial issue. It involves changing the thin, cold atmosphere – which we cannot breathe – into a warmer climate with air that humans can breathe outdoors. It would require international agreement, and it would take centuries to complete the first step; and there are also religious and ethical issues that must be considered. What if there is already simple life on Mars that would not survive in an oxygen-rich atmosphere? Would such alterations to the conditions be permanent, or is Mars simply incapable of sustaining a life-supporting atmosphere? Can or should humans play God?

The 'space age' envisioned during the 1950s and 1960s is far removed from the attitudes about space today. The vision of space being there to be 'conquered' by 'man' is a quaint historical perspective in the modern world – as much a product of the time as Wernher von Braun's fleet of spaceships in his Mars Project, or the Dan Dare stories.

A modern perspective is provided by Carl Sagan, quoted in Michael Huang's 'Sagan's rationale for human spaceflight': 'If we were up there among the planets, if there were self-sufficient human communities on many worlds, our species would be insulated from catastrophe ... A cataclysmic impact on one world would likely leave all the others untouched. The more of us beyond the Earth, the greater the diversity of worlds we inhabit ... then the safer the human species will be.'

Marswalk One will be a defining moment in human history. Beckoning on the horizon would be more expeditions to Mars, leading to bases, possibly colonies, and perhaps, in the distant future, terraforming of the planet. How much of this will actually happen will depend upon political will, public interest, and the results of the first human exploration. Will we seize the chance to step 'beyond the cradle', or will we choose to go the way of Apollo and stop at the height of our achievement? Time, as always, will tell.

**REFERENCES**

1  Benton C. Clark, *Crew Activities, Science and Hazards of Manned Missions to Mars*, Planetary Sciences Laboratory, Martin Marietta Astronautics, Denver, Colorado, 39th Congress IAF, Bangalore, India, 8–15 October 1988, IAF-88-403.
2  Russell Schweickart *et al.*, 'The asteroid tugboat', *Scientific American*, **289**, no. 5, November 2003.
3  *Planetary Exploration through Year 2000: an Augmented Program*, NASA SSEC.
4  *Orbiting Quarantine Facility: the Antaeus Report*, NASA SP-454
5  *Biological Contamination of Mars: Issues and Recommendations*, Space Studies Board, National Research Council.
6  Dubeck, Moshier and Boss, *Fantastic Voyages: Learning Science through Science Fiction Films*, American Institute of Physics.
7  Wheeler, *Antarctica, the Falklands and South Georgia.*

# Chronology

| | | |
|---|---|---|
| BC | | Mars recorded by the Egyptians, Chinese, Babylonians, Greeks, Assyrians, and others. |
| AD 1576 | | Naked-eye measurements of the position of Mars in the night sky. |
| 1659 | | Christiaan Huygens discovers dark markings on the surface of Mars, and helps determine the length of the martian day. |
| 1877 | | Asaph Hall discovers the two moons of Mars: Phobos and Deimos. The measurement of their orbits helped determine the mass and surface gravity of the planet. In the same year, Italian astronomer Giovanni Schiaparelli announces his discovery of *canali* – 'channels', and the misinterpretation of the word as 'canals' leads to 'Mars mania'. |
| 1894 | | Percival Lowell interprets his drawing of the 'canals' as an irrigation network – 'a clear indication of intelligence on the surface'. |
| 1896 | | Percival Lowell writes: 'That Mars is inhabited by beings of some sort or other is as certain as it is uncertain what these beings may be.' |
| 1948 | | Wernher von Braun completes an unpublished novel about an expedition to Mars. |
| 1952 | | Wernher von Braun's *Das Marsprojekt* is published in West Germany, and in the following year is translated into English – *The Mars Project*. |
| 1954 | Apr 30 | *Collier's* publishes an article that describes the exploration of Mars, based on von Braun's *The Mars Project*. |
| 1956 | | Wernher von Braun and Willey Ley co-author *The Exploration of Mars*, based on the *Collier's* articles. |
| | Dec 4 | Walt Disney premiers the TV programme *Mars and Beyond*, featuring nuclear-electric-powered spacecraft. |
| 1958 | | Sergei Korolyov and Mikhail Tikhonravov prepare a detailed document listing the goals of the Soviet space programme |

|      |          | based on the vision of Konstantin Tsiolkovsky. It includes human flights to Mars. |
|------|----------|---|
| 1959 | Apr | NASA Lewis Research Center staff testify to Congress, requesting funding for a Mars expedition study during Fiscal Year 1960. |
| 1960 | Jun | A Soviet Government Decree on the future of Soviet space exploration includes Sergei Korolyov's plans for sending cosmonauts to fly around Mars and back to Earth again. |
|      | Oct 10 | The Soviet Union attempts to send an unmanned spaceprobe towards Mars, but it fails to reach Earth orbit. Four days later, a second probe fails to launch. These would have been the first planetary probes. |
| 1961 | Oct 12 | OKB-1 designers complete the initial technical calculations for the Heavy Interplanetary Ship-1 (TMK) for human flight to Mars. |
| 1962 |  | Ernst Stuhlinger's group at MSFC proposes a piloted mission concept for launch in the early 1980s. |
|      | May | MSFC Future Projects Office begins manned Mars mission studies under the Early Manned Planetary–Interplanetary Roundtrip Expeditions (EMPIRE) project. |
|      | Oct 24 | The Soviet Union attempts to send a probe to fly by Mars, but it fails to leave Earth orbit. |
|      | Nov 1 | The Soviet Union's Mars 1 becomes the first probe to successfully leave Earth orbit and travel towards Mars for fly-by; contact is lost on 21 March 1963, and it is estimated to have flown past Mars at 193,000 km on 19 June 1963. |
|      | Nov 4 | The Soviet Union attempts to send a probe to land on Mars, but it fails to leave Earth orbit. |
| 1963 | May 21–23 | A two-day Manned Planetary Mission Technical Conference, held at NASA's Lewis Research Center, includes papers on the design of a Mars Excursion Module. |
| 1964 | Jun 21 | TRW space technology, under contract from NASA Ames Research Center, presents the Manned Mars Mission Study. |
|      | Sep | The Unfavourable Mars Manned Planetary–Interplanetary Roundtrip Expeditions (UMMPIRE) study is initiated by the Marshall Future Projects Office and contracted to Douglas Aircraft Company and General Dynamics. |
|      | Nov 5 | Mariner 3 is lost when the launch vehicle fails. Mariner 4 is successfully launched on 28 November. On 30 November the Soviets launch Zond 2 to Mars, but contact lost in April 1965. On 6 August 1965 it flew by Mars at 1,500 km. |
| 1965 | Feb | NASA publishes the Manned Planetary Reconnaissance mission, by Harry Ruppe, of the MSFC Future Projects Office, which evaluated the use of adapted Apollo–Saturn hardware to perform fly-bys of Mars and Venus. |

|      |        |                                                                                                                                                                                                                                                                                                                                                                                                                                                                                                                                                                                                                                                                                                                       |
|------|--------|---|
|      | Jul 15 | Mariner 4 becomes the first successful Mars fly-by spacecraft on closest approach (9,600 km), and takes twenty-one photographs of the surface. |
| 1966 |        | The Planetary Joint Action Group (JAG) evaluates existing and near-term technology to mount manned Mars fly-by missions, leading to Mars landing missions. |
| 1967 | Aug 22 | The US House cuts $500 million from the NASA Fiscal Year 1968 budget, effectively cancelling the Voyager unmanned Mars Landers and any chance of a 1970s manned Mars mission. |
| 1968 | Jan    | Boeing publishes its final report on a 14-month NASA contract study into nuclear spacecraft, which in part would support manned Mars landings. |
| 1969 | Jan 25 | Soviet Chief Designer V.P. Mishin meets with Chief Designer Pilyugin to discuss a three-step N-1-based Soviet Mars Exploration programme that includes sample-return missions in 1973, a Mars manned orbital mission in 1975, and a manned Mars landing by 1977. During a second meeting four days later, there is a suggestion for a Multi-role Space Base Station (MKBS) to mount a manned Mars expedition. The 1970s would see an advanced and sophisticated Soviet Mars programme. |
|      | Mar 27 | The Soviets attempt the first launch of a dual-launch Mars mission, with the second launch attempted on 4 April. Both of them suffer booster failure. |
|      | Jul 31 | Eleven days after the Apollo 11 lunar landing, Mariner 6 completes its closest approach to Mars at 3,429 km, and taking twenty-five photographs (fifty additional images were taken during approach). |
|      | Aug 4  | Wernher von Braun and NASA Administrator Tom Paine present a proposal to the Space Task Group for a manned Mars Expedition in 1982. |
|      | Aug 5  | Mariner 7 flies its closest approach to Mars at 3,430 km. It takes ninety-three far-encounter and thirty-three close-up photographs. |
|      | Sep 15 | The STG report on the post-Apollo space programme for America fails to support an aggressive Mars exploration programme. |
| 1970 | Mar 7  | President Nixon reveals his post-Apollo space programme policy, which includes the statement: 'We will eventually send men to explore the planet Mars.' |
| 1971 | Feb    | NASA publishes its manned Mars mission designs incorporating Shuttle technology and chemical propulsion. It will be more than a decade before NASA publishes another manned Mars mission design. |

| | | |
|---|---|---|
| | May 8 | Mariner 8 is lost in a failure of the launch vehicle. Mariner 9 launches on 30 May and is successfully dispatched toward Mars. During May 1971 the USSR dispatches three spacecraft to Mars: Cosmos 419 (10 May) fails to leave Earth orbit, but Mars 2 (19 May) and Mars 3 (28 May) head for Mars. |
| | Nov 14 | Mariner 9 becomes the first spacecraft to enter Mars orbit. Over the next year it transmits more than 7,300 images of the planet and its moons. It arrived during a planet-encircling dust-storm that did not clear until January 1972, and seriously affected the performance of the Soviet Mars 2 and Mars 3 craft. |
| | Nov 27 | The Mars 2 lander crashes on the surface, but becomes the first spacecraft to 'land' on the planet. The orbiter enters orbit and returns twelve images. |
| | Dec 2 | Mars 3 achieves the first soft landing on the surface of Mars. It returns 20 seconds of TV signals (without any contrast) before contact is lost. The orbiter enters orbit, and its images reveal thick clouds of dust. |
| 1974 | Feb–Mar | A 'fleet' of four USSR Mars spacecraft arrives at Mars after being dispatched during July and August 1973. Mars 4 and Mars 5 are orbiter craft. Mars 4's main engine fails, preventing it from entering orbit, but its imaging system returns some pictures during close approach (2,200 km) on 10 February 1974. Mars 5 enters orbit on 12 February, and later acts as a relay for Mars 6 and Mars 7. It takes up to sixty images, and measurements of the planet. Mars 6 and Mars 7 incorporate a lander and a fly-by section. The Mars 6 lander touches down on 12 March. Parachutes are deployed, and some data is retrieved up to a few seconds before landing, when contact is lost. Mars 7 ejects its lander on 9 March, but misses the planet by 1,300 km, indicating a failure with the onboard braking or attitude control systems. |
| 1976 | Jul 20 | The high point of the American bicentennial celebrations is the landing of Viking 1 on Mars (exactly 7 years after Apollo 11 landed on the Moon). The vehicle had arrived in orbit on 19 June. The second Viking enters orbit on 7 August, and its lander touches down on 3 September. Through to 1982 the landers return a total of more than 4,500 pictures from the surface, and the orbiters transmit 51,539 images, mapping 97% of the surface to 300-metre resolution and 2% to 25 metres or better. |
| 1978 | | Martin Marietta publishes the results of the Viking missions, including an exploration of the life support implications of the results, entitled *The Case for Man on Mars*. |
| 1981 | | The first Case for Mars conference is held in Boulder, Colorado. |

| 1983 | | The most detailed piloted Mars mission study in twelve years is commissioned by the Planetary Society. |
| 1984 | Jul | Case for Mars II conference. |
| 1986 | Jun | NASA's Johnson Space Center proposes a response to suspected Soviet manned Mars plans by utilising elements of space-station and lunar-base hardware to support a US Mars fly-by mission in the 1990s. |
| 1987 | Apr | The US and the USSR review the Space Cooperation Agreement, initiating calls for joint studies to send international expeditions to the Mars. |
| | Aug | NASA publishes its Leadership and America's Future in Space report, indicating extended and expanded exploration of Mars |
| 1988 | | *Beyond Earth's Boundaries: Human Exploration of the Solar System in the Twenty-First Century* is published by NASA to help define and direct future long-term space planning in the new century. |
| | Aug 31 | Phobos 1 – launched on 7 July 1988 – is sent an erroneous radio signal from the flight control centre in Kaliningrad, Moscow region, causing it to lose attitude control and power. |
| 1989 | | Case for Mars III conference. The 90-day Study into Space Exploration Initiatives is published. |
| | Jan 29 | Phobos 2 – launched on 12 July 1988 – enters Mars orbit, but fails prior to its close approach to Phobos. |
| | May 31 | *EVA in Mars Surface Exploration: Final Report* is published as part of NASA's Advanced EVA Systems Requirements Definition Study. |
| 1990 | | Robert Zubrin and David Baker propose the Mars Direct approach for early manned Mars missions in the current decade. Case for Mars IV conference. |
| 1991 | Jul | *Exploring the Moon and Mars: Choices for the Nation* is published by the US Congress Office of Technical Assessment to determine robotic and automated technologies to support lunar and Mars exploration as part of the new goals of the US space programme. |
| 1993 | | The Mars Exploration Study Team – a Synthesis Group of SEI – publishes its findings. |
| | Aug 21 | Contact with Mars Observer – launched on 25 September 1992 – is lost immediately prior to arriving at Mars. |
| 1996 | Nov 16 | A launch vehicle failure on the Russian Mars-96 mission loses the orbiter and four landers. |
| 1997 | Jul 4 | Mars Pathfinder – launched 14 December 1996 – lands on Mars, and Sojourner becomes the first rover to operate on Mars. |

|        | Sep 12 | Mars Global Surveyor – launched on 7 November 1996 – arrives in orbit to complete a very-high-detail orbital mapping mission through to January 2000, and an extended second mission through to the end of 2006. |
| 1998 | Jul 4 | Japan becomes third nation to launch a probe towards Mars. Nozomi arrives at Mars in December 2003, delayed due to problems with its propulsion system. Its frozen propellant tanks could not be thawed to allow it to enter Mars orbit |
| 1999 | Sep 23 | Contact with Mars Climate Orbiter – launched on 11 December 1998 – is lost when it arrives at Mars. |
|        | Dec 3 | Mars Polar Lander and two soil probes – launched on 3 December 1999 – are lost on arrival at the south polar region. |
| 2001 | Feb | NASA publishes the *Humans to Mars* monograph by David S.F. Portree, documenting fifty years of human intention to explore Mars, from early mission planning to the year 2000. |
|        | Oct 24 | Mars Odyssey – launched on 17 March 2001 – enters orbit to conduct its primary mission involving global composition, ground ice and thermal imaging. |
| 2003 | Dec 25 | Mars Express – ESA's first interplanetary mission – arrives in Mars orbit. Contact with the Beagle 2 lander (UK) is lost as it descends to the surface of the planet. |
| 2004 | Jan 4 | The first Mars Exploration Rover mission lands on the surface with the Spirit rover intact. |
|        | Jan 14 | President George W. Bush announces a major redirection for the US space programme, including a return to the Moon by 2015, perhaps leading to human flights to Mars by around 2030. |
|        | Jan 25 | The second Mars Exploration Rover mission arrives with the Opportunity rover. |

The above summary clearly indicates the difficulty of landing on Mars, and why only a handful of payloads have reached the surface (and not always in working order).

## Landing site coordinates

| Spacecraft | Country | Date | Coordinates | Result |
|---|---|---|---|---|
| Mars 2 | USSR | 1971 Nov 27 | 45° S, 58° E | Crashed |
| Mars 3 | USSR | 1971 Dec 2 | 45° S, 158° W | Soft landing |
| Mars 6 | USSR | 1974 Mar 12 | 24° S, 25° W | Soft landing? |
| Viking 1 | USA | 1976 Jul 20 | 22°.48, N 47°.94 | Soft landing |
| Viking 2 | USA | 1976 Sep 3 | 47°.96, N 225°.71 W | Soft landing |
| Mars Pathfinder | USA | 1996 Dec 14 | 19°.30, N 33°.52 W | Soft landing |
| Mars Polar Lander | USA | 1999 Dec 3 | 76° S, 195° W* | Crashed |
| Amundsen (DS-2) | USA | 1999 Dec 3 | 73° S, 210° W* | Crashed? |
| Scott (DS-2) | USA | 1999 Dec 3 | 73° S, 210° W* | Crashed? |
| Beagle 2 | UK/ESA | 2003 Dec 25 | 11°.6 N, 269°.5 W* | Crashed? |
| Spirit Rover | USA | 2004 Jan 5 | 14°.82 S, 184°.85 W* | Soft landing |
| Opportunity Rover | USA | 2004 Jan 25 | 2°.07 S, 6°.08 E* | Soft landing |

* Centre of landing ellipse

## Future missions

| 2005 | NASA | Mars Reconnaissance Orbiter |
|---|---|---|
| 2007 | NASA | Phoenix |
| 2009 | NASA | Mars Telecommunications Orbiter |
| | NASA | Mars Science Laboratory (smart lander) |
| | ESA | Exomars |
| 2011 | ESA | Sample return mission |
| 2014 | NASA | Sample return mission |

# Bibliography

Prior to 2004 there was little reference material about the initial surface exploration of Mars, but a wealth about the planet itself – telescopic observations, robotic exploration, and the prospect of extended human exploration and terraforming of the planet. The following are the main sources of references concerning what may or may not be accomplished on the very first EVA periods on the surface of Mars. In the compilation of this book, the authors also sought the advice of a number of individuals who have examined the potential of human flights to Mars.

**Interviews**

Dr Charles Frankel, geologist and F. Mars occupant, 9 January 2002

Dr David Rothery, geologist, Open University, 6 August 2002

Dr Alex Ellery, senior lecturer and robot technologist, Kingston University, 10 October 2003 and 4 February 2004

David A. Hardy, space artist, 20 November 2003

Bo Maxwell, Chairman of the UK Mars Society and MDRS occupant, 5 and 16 December 2003

Graham Worton, field geologist and former senior environmental geologist, Johnson-Poole-Bloomer, 4 December 2003

Dr Ian Crawford, geologist, Department of Earth Sciences, Birkbeck College, UCL, 15 January 2004

Dr Michael Duke, geologist, former Mars mission planner, Colorado School of Mines, 28 January 2004

**Books and reports**

1970    *Space Annual*, Peter Fairley, TV Times Publications, London.

1971    *Pioneering in Outer Space*, Hermann Bondi *et al.* Heinemann Educational Books.

1984    *On Mars: Exploration of the Red Planet, 1958–1978*, E.C. and L.N. Ezell, NASA SP-4212.

1987    *Leadership and America's Future in Space*, A Report to the Administrator, Sally K. Ride, NASA.

*Solar System Log*, Andrew Wilson, Jane's Publishing Company.

1988   *Race to Mars: The ITN Mars Flight Atlas, ed.* Frank Miles and Nicholas Booth.

*On Mars*, P.A. Moore, Cassell.

*Beyond Earth's Boundaries: Human Exploration of the Solar System in the Twenty-First Century*, Annual Report to the Administrator, NASA Office of Exploration.

1989   *Extravehicular Activity in Mars Surface Exploration, Final Report, 31 May 1989*, part of the Advanced Extravehicular Activity Systems Requirement Definition Study prepared for NASA (NAS9-17779, Phase III).

*Report of the 90-Day Study on the Human Exploration of the Moon and Mars*, NASA.

1990   *Mission to Mars*, Michael Collins, Grove Weidenfeld.

*Design Considerations for Future Planetary Space Suits*, Joseph J. Kosmo, NASA LBJ Space Center, SAE Technical Paper 091428, presented at the 20th Intersociety Conference on Environmental Systems, Williamsburg, Virginia, 9–12 July 1990.

1991   *Exploring the Moon and Mars: Choices for the Nation*, US Congress Office of Technology Assessment, OTA-ISC-502, July 1991.

1997   *The Case for Mars*, Robert Zubrin, Touchstone Books.

1998   *Uncovering the Secrets of the Red Planet*, P. Raeburn, National Geographical Society.

*The Viking Mission to Mars*, Martin-Marietta, Library of Congress No. 75-16764.

1999   *The New Solar System* (fourth edition), Cambridge University Press.

*Extravehicular Activity Suit Systems Design: How to Walk, Talk and Breathe on Mars*, Cornell University Press.

2000   *Challenge to Apollo*, Asif Siddiqi, NASA SP-2000-4408.

*Towards Mars*, ed. Risto Pellinen and Paul Raudsepp, Rand Publishing.

2001   *The Origins and Technology of the Advanced EVA Space Suit*, Gary L. Harris, American Astronautical Society, AAS History Series, Vol. 24.

*Humans to Mars: Fifty Years of Mission Planning, 1950–2000*, David S.F. Portree, NASA Monographs in Aerospace History, No. 21, NASA SP-2001-4521.

*Technologies for Exploration, Aurora Proposal: Annex D*, ed. Andrew Wilson, ESA SP-1254.

2003   *Russian Spacesuits*, Isaak P. Abramov and Å. Ingemar Skoog, Springer–Praxis.

*Some Problems of Selection and Evaluation of the Martian Suit Enclosure Concept*, I. Abramov, N. Moiseyev and A. Stoklitsky, Zvezda, Russia, presented at the 54th IAF Congress, 29 September–3 October 2003, Bremen, Germany (IAC-03-IAA.13.3.02).

*Design of an EVA Suit Suitable for use on the Martian Surface*, Laura Parker, Cranfield University, UK; presented at the 54th IAF Congress, 29 September–3 October 2003, Bremen, Germany (IAC-03-IAA.13.3.03).

2002   *Mapping Mars*, O. Morton, Fourth Estate.

**Web sites**
BBC News, Science and Nature section, http://www.bbc.co.uk/news
Beagle 2, http://www.beagle2.com
European Space Agency, http://www.esa.int
Mars Society, http://www.marssociety.org.uk
NASA Center for Mars Exploration, http://cmex-www.arc.nasa.gov
National Space Science Data Center, http://nssdc.gsfc.nasa.gov
Natural History Museum, London, http://www.nhm.ac.uk
Romance to Reality (David Portree), http://rtr.marsinstitute.info
NASA's Mars Exploration Program, http://mars.jpl.nasa.gov
First Landing Site Workshop for the 2003 Mars Exploration Rovers
   http://www.lpi.usra.edu/meetings/mer2003/pdf/program.pdf

# Index